CLIMATE-RESILIENT FISCAL MANAGEMENT

EXPERIENCE FROM SOUTHEAST ASIA

NOVEMBER 2024

Contents

Figures and Boxes

FIGURES

BOXES

Foreword

Southeast Asia is at a critical juncture in addressing climate risks, which are rapidly intensifying and imposing significant burdens on public finances. Finance and planning ministries are facing unprecedented exposures to these risks. From extreme weather events to the economic pressures of transitioning to low-carbon economies, the demands on fiscal systems are mounting, requiring urgent and strategic action.

This report, *Climate-Resilient Fiscal Management: Experience from Southeast Asia*, provides a timely blueprint to help governments face these challenges. It consolidates earlier work by the Asian Development Bank (ADB) and the Coalition of Finance Ministers for Climate Action (CFMCA) in 2023 and presents a framework to assess and manage climate-related fiscal risks and mobilize climate finance, drawing on good practices and lessons learned across the region. By embedding climate risks into fiscal planning and management efforts, finance and planning ministries can strengthen their ability to adapt to evolving risks while contributing to broader decarbonization and climate resilience goals.

Key insights from the report highlight the growing volatility in public finances caused by physical climate risks, which are already highly material and expected to escalate over the coming decades. Furthermore, the report stresses that the region's heavy reliance on emission-intensive sectors will create significant fiscal pressures as the global economy shifts toward cleaner energy. To navigate these challenges, the report identifies actionable strategies to assess and manage climate-related fiscal risks and mobilize public and private capital to support investments in clean energy and climate-resilient infrastructure.

The report presents a clear framework in three steps: assessing and embedding climate risks in fiscal planning, strategically managing these risks through policy reforms and risk-sharing mechanisms, and mobilizing the necessary capital to fund adaptation and mitigation efforts. Further, the report identifies short- and medium-term priorities to help finance and planning ministries adapt to their respective stages of development in adopting these practices, offering tailored guidance for immediate action.

As the urgency of climate adaptation and mitigation continues to grow, the insights and practical tools in this report are vital for decision-makers across Southeast Asia. By leveraging these strategies, finance and planning ministries can not only safeguard their fiscal stability but also foster sustainable economic growth in the face of a changing climate.

We hope this report provides valuable guidance for finance and planning ministries as they lead the way in building a climate-resilient future for Southeast Asia.

Bruno Carrasco
Director General
Climate Change and Sustainable
  Development Department

Ramesh Subramaniam
Director General and Group Chief
Sectors Group

Winfried F. Wicklein
Director General
Southeast Asia Regional Department

Acknowledgments

This report is a knowledge product of the Asian Development Bank (ADB) with funding support from the technical assistance for Accelerating Climate Transitions through Green Finance in Southeast Asia.

The report is a result of the team effort led by Heejin Lee (climate change specialist, Climate Change and Sustainable Development Department [CCSD]), with support from Naeeda Crishna Morgado (senior infrastructure specialist (climate finance), Southeast Asia Regional Department [SERD]), Nanki Kaur (senior climate change specialist (climate change adaptation), CCSD), Luis Almeida (public management specialist, Sectors Group [SG]), Brendan A. Coleman (climate change specialist, CCSD), Swati Mitchelle Dsouza (climate change specialist (just transition), CCSD), Laisiasa Tora (principal public sector specialist, SG), and Sharmaine Dianne G. Ramirez (consultant, SERD).

The report draws on research and analysis from McKinsey & Company, led by Oliver Walker, with support from Alina Steinweg, Jinyi Zhou, and Lottie Plaschkes. It benefited significantly from inputs by technical and country experts, including Benita Ainabe (principal financial sector specialist, SG), Emma R. Allen (senior country economist, SERD), Marina Lopez Andrich (contractor, SERD), Binh H. Bui (climate change officer, CCSD), Meutia Chaerani (senior climate change specialist, SG), Aekapol Chongvilaivan (senior planning and policy specialist, Strategy and Policy Department), Vu Quang Dang (consultant, SG), Virender Kumar Duggal (principal climate change specialist, CCSD), Andrew Fransciscus (senior programs officer, SERD), Anna M. Fink (senior country economist, SERD), Sree Kartha (consultant, SERD), Gary Krishnan (senior country specialist, SERD), Cristina Lozano (principal country specialist, SERD), Anouj Mehta (country director, SERD), Chu Hong Minh (senior financial sector officer, SG), Sophie Duong Nguyen (senior country economist, SERD), Nguyen Ba Hung (principal country economist, SERD), Chameroen Ouch (senior programs officer (governance), SG), Maria Joao M. Pateguana (principal financial management specialist, Procurement, Portfolio, and Financial Management Department), Alfredo Perdiguero (regional head, Regional Cooperation and Integration, SERD), Ghia V. Rabanal (associate climate change officer, CCSD), Ahmad Faris Rabidin (principal financial sector specialist, SG), Lazeena Rahman (senior climate finance specialist, SG), Ron H. Slangen (deputy country director, SERD), Thavin So (senior climate change officer, CCSD), Jackie B. Surtani (country director, SERD), Joel Syquia (principal public sector specialist, SG), Sonomi Tanaka (country director, SERD), Jhelum T. Thomas (principal public sector specialist, SG), Jiro Tominaga (country director, SERD), Jyotsana Varma (country director, SERD), and Marianne O. Villanueva (associate operations coordination officer, CCSD).

Guidance was also received from Bruno Carrasco (director general, CCSD), Toru Kubo (senior director, CCSD), Tariq Niazi (senior director, Public Sector Management and Governance, SG), Noelle Obrien (director, CCSD), Scott Roberts (head, Green Finance Hub Unit, SERD), Ramesh Subramaniam (director general and group chief, SG), and Winfried F. Wicklein (director general, SERD).

The conclusions and recommendations in this report do not necessarily reflect the views of any of the contributors.

Overall production coordination of the report was managed by Criselda G. Rufino (Infrastructure Officer, SERD). Editing was done by Layla Tanjutco-Amar; design and layout by Rocilyn Laccay; proofreading by Jess Alfonso A. Macasaet; and page proof checking by Ma. Cecilia Abellar.

Abbreviations

ACGF	–	ASEAN Catalytic Green Finance Facility
ADB	–	Asian Development Bank
ASEAN	–	Association of Southeast Asian Nations
CBAM	–	Carbon Border Adjustment Mechanism
CFA	–	central finance agency
CFMCA	–	Coalition of Finance Ministers for Climate Action
CFPP	–	coal-fired power plant
C-PIMA	–	Climate-Public Investment Management Assessment
DFI	–	development financial institute
EV	–	electric vehicle
GDP	–	gross domestic product
GPP	–	green public procurement
GSB	–	Government Savings Bank
IFI	–	international financial institution
IMF	–	International Monetary Fund
IPCC	–	Intergovernmental Panel on Climate Change
Lao PDR	–	Lao People's Democratic Republic
LPG	–	liquefied petroleum gas
LTS	–	long-term climate strategies
MAS	–	Monetary Authority of Singapore
MDB	–	multilateral development bank
MTBF	–	medium-term budget framework
MTFF	–	medium-term fiscal framework
NAP	–	national adaptation plan
NDC	–	nationally determined contributions
NGFS	–	Network for Greening the Financial System
OECD	–	Organisation for Economic Co-operation and Development
PDNA	–	post-disaster needs assessment
PFB	–	Pooling Fund Bencana (Pooling Fund for Disasters)
PFM	–	public financial management
PIM	–	public investment management
PIMA	–	Public Investment Management Assessment
PPP	–	public–private partnership
SDG	–	Sustainable Development Goal
SEADRIF	–	Southeast Asia Disaster Risk Insurance Facility
SMEs	–	small and medium-sized enterprises
SOB	–	state-owned bank
SOE	–	state-owned enterprise
SWF	–	sovereign wealth fund
VCM	–	voluntary carbon market

Executive Summary

Acting on climate risks and opportunities is a key priority for finance and planning ministries across Southeast Asia. These ministries, responsible for government financial functions, including budget and investment management, tax and subsidy policy, debt management and financing, monitoring of public entities, and financial system regulation, are referred to as central finance agencies (CFAs) in this report. CFAs in Southeast Asia face significant and rapidly growing exposures to climate risks.

- **Increasing physical risks can cause volatility in public finances.** Acute physical risks are already highly material—relief and recovery costs alone for climate-related disasters can reach 20% of government budgets in the region—and are projected to grow rapidly to the middle of the century, exacerbated by chronic risks such as sea level rise.

- **Transition risks will create significant fiscal pressures.** Public finances in Southeat Asia are strongly dependent on emission-intensive sectors, with numerous large state-owned oil and gas firms and many governments deriving more than 15% of revenues from fossil fuel-linked sources.

- **CFAs play a key role in mobilizing the investment** required for broader adaptation and decarbonization programs. Estimated financing needs for clean energy alone is $150 billion per year and financing needs for adaptation are even greater at $387 billion per year, far exceeding the current level of international public finance.

This report, *Climate-Resilient Fiscal Management: Experience from Southeast Asia,* considers how CFAs can respond strategically to climate risk, highlighting good practices from the region and identifying crosscutting priorities. It consolidates earlier work by the Asian Development Bank (ADB) and the Coalition of Finance Ministers for Climate Action (CFMCA) in 2023 and presents a framework to assess and manage climate-related fiscal risks and mobilize climate finance, illustrated by Figure E1:

- **Assessing and embedding climate risk implications in fiscal planning frameworks.** This requires translating evidence from national and international sources on physical and transition risks to implications for CFAs and deploying modeling tools to quantify key macrofiscal impacts. The risk assessment can then be embedded in decision-making on investment planning and budget management, and potentially disclosed to investors.

- **Strategically managing climate-related fiscal risks.** Actions here span risk assignment, reduction, and transfer. First, by assigning "risk ownership," CFAs can contain liabilities and reduce moral hazard related to risks faced by state-owned enterprises (SOEs), local governments, and the private sector. Second, CFAs can drive strategic approaches to investment in climate resilience and decarbonization goals to reduce risks, ensuring line ministries take consistent approaches to planning, procurement, and financing. Third, CFAs can change tax and spending regimes to meet the twin objectives of providing incentives for adaptation and mitigation projects and of ensuring long-term fiscal sustainability: interventions such as carbon pricing and phasing out fossil fuel subsidies can potentially support both goals. Finally, disaster risk finance strategies can improve CFAs' fiscal stability, while enhancing their capacity to respond to climate-related disasters.

- **Mobilizing public and private capital.** CFAs can take a proactive role in securing finance for climate-related investment. This includes creating capacity for public spending, leveraging climate finance instruments, raising revenues through carbon markets, and through the use of other innovative approaches such as debt-for-nature swaps. To attract private capital, CFAs can work with development finance institutions (DFIs) and international partnerships such as the ASEAN Catalytic Green Finance Facility (ACGF) to develop pipelines of projects, with these entities acting as anchor investors to catalyze investor participation at scale. The domestic financial sector can be incentivized to prioritize climate finance through the adoption of green taxonomies or climate risk disclosure frameworks.

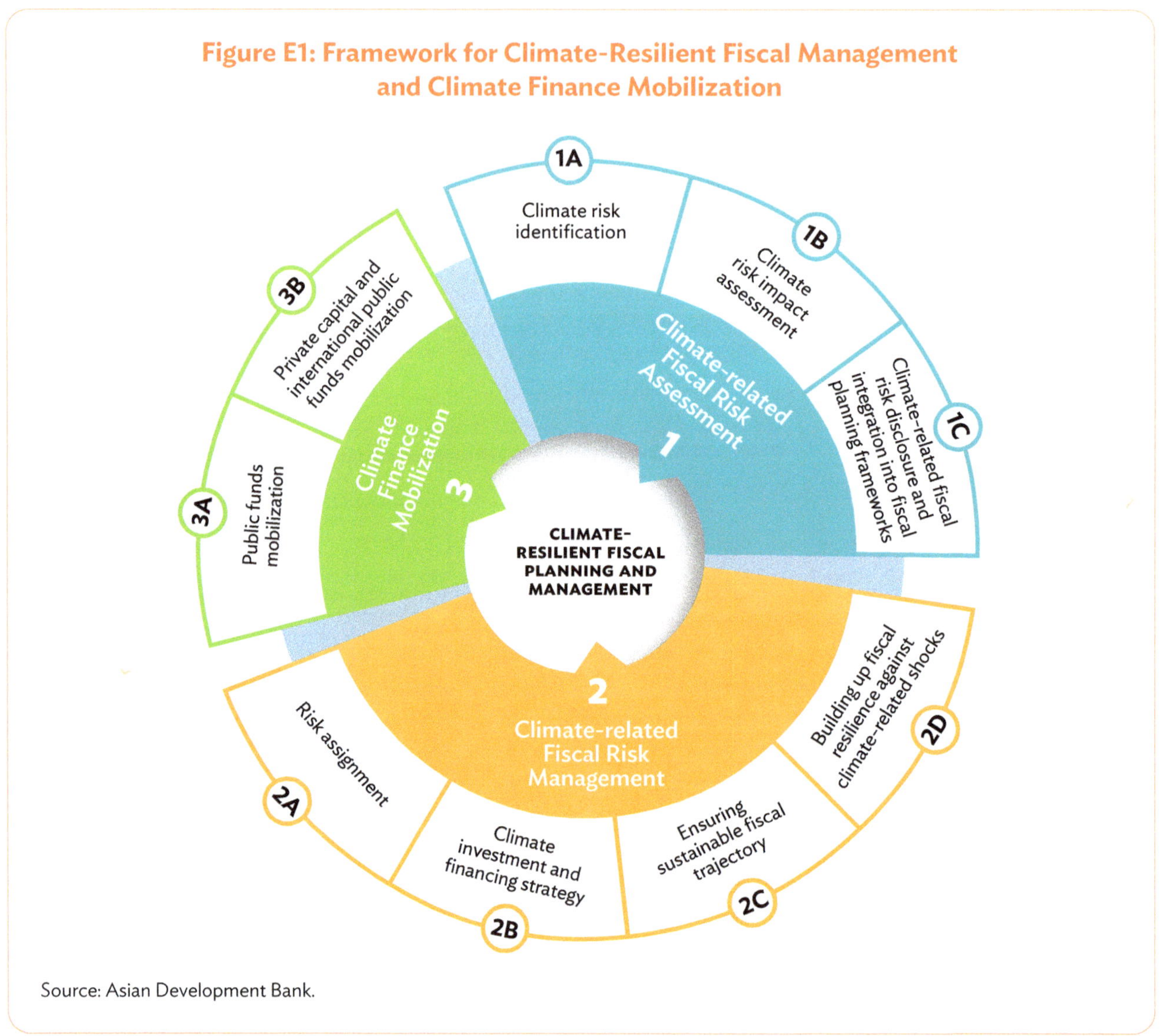

Figure E1: Framework for Climate-Resilient Fiscal Management and Climate Finance Mobilization

Source: Asian Development Bank.

Short- and medium-term priorities for CFAs in Southeast Asia will depend in part on how advanced they are on the journey of adopting good practices. Figure E2 highlights country-specific examples of good practices and Figure E3 identifies key opportunities for climate-resilient fiscal management and climate finance mobilization. While all the steps of the framework can be pursued synergistically, CFAs may prioritize interventions based on the degree to which good practices are already established, as well as countries' specific exposures to physical and transition risks.

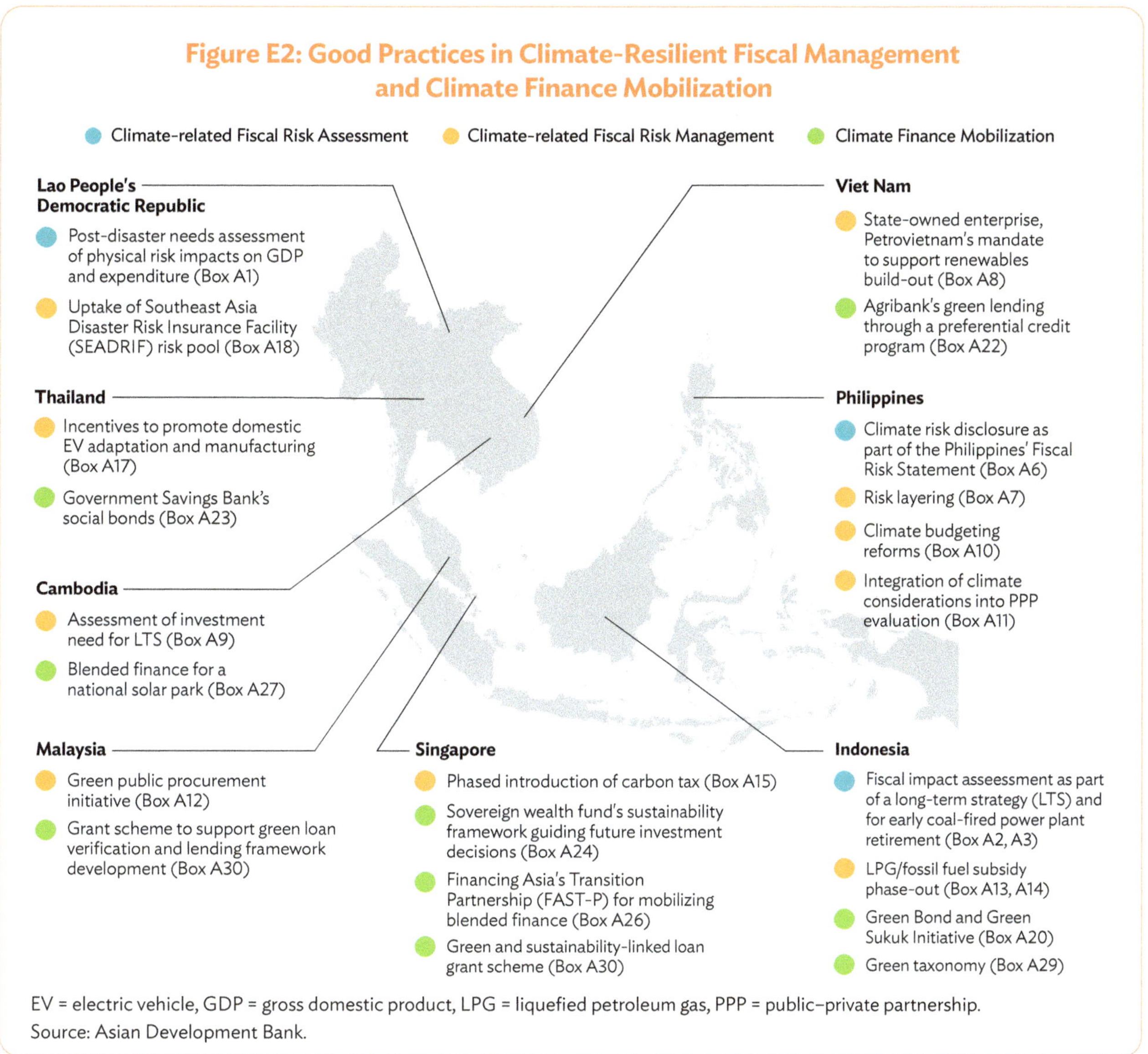

EV = electric vehicle, GDP = gross domestic product, LPG = liquefied petroleum gas, PPP = public–private partnership.
Source: Asian Development Bank.

Speaking generally across Southeast Asia:

- **An initial step for all CFAs is to avoid short-term harm while building capacity to take a more strategic approach.** Given the Southeast Asia's high exposure to physical risks, a short-term priority is to reduce the destabilizing effects of climate disasters on public finances and the broader economy through the adoption of disaster risk finance strategies founded on risk layering. This requires clarity of risk ownership between government agencies and the private sector, informed by targeted modeling of the link between physical risks and fiscal conditions. CFAs can look to disaster risk financing and management approaches adopted in the Philippines, which has contingent access to millions of dollars to support relief and recovery, and the Lao People's Democratic Republic (the Lao PDR), which uses a portfolio of disaster insurance instruments through the Southeast Asia Disaster Risk Insurance Facility (SEADRIF) initiative. Another way to circumvent harm is by ensuring planned investment projects do not lock in high emissions or vulnerability to climate risk. CFAs can assess sector programs for climate risk, including fiscal implications, and adjust course where required. Over the medium term, these capabilities can evolve to incorporate a more comprehensive understanding of climate-related fiscal risks into investment and finance strategies for low-carbon, resilient development. For credit-constrained nations, an effective approach may be to engage with DFIs and donors to secure concessional finance to support these programs, while others may mobilize state resources such as sovereign wealth funds.

- **CFAs with more established approaches can focus on supporting economy-wide climate resilience and decarbonization goals.** Such CFAs may already understand the macrofiscal importance of decarbonization and resilience objectives at a high level and have formulated sector investment plans. A short-term priority for these CFAs is to develop balanced fiscal pathways that support these goals, potentially involving the phased implementation of carbon pricing or subsidy reforms, with measures to protect vulnerable groups. In Indonesia, for example, fuel subsidy reforms have provided targeted disbursements worth $30 to low-income households. A further priority, especially given limited public borrowing capacity, is to develop credible financing strategies for investment plans by partnering with DFIs and the private sector to effectively deploy blended instruments at scale. The cumulative effect of these interventions on fiscal sustainability and stability can be monitored over time using quantitative modeling under a range of scenarios.

- **Finally, the most advanced CFAs can move best practices forward.** Their focus may include more advanced modeling of the relationship between climate risks and fiscal outcomes and the development of new revenue models that can attract private finance for adaptation projects. By developing advanced tools that can be applied elsewhere, these CFAs can play a catalytic role across the region. Singapore, for example, has established the Financing Asia's Transition Partnership (FAST-P) aimed at mobilizing $5 billion for investment to address both local and regional needs.

To move forward, CFAs can use the framework to set priorities and develop action plans. An initial assessment can reveal gaps between current policy frameworks and good practices outlined in this report. This will help shape strategic priorities, enabling CFAs to design reform programs and identify the resources required to implement them. For many CFAs, these resources may include partnerships with organizations like ADB that can offer funding, technical assistance, and capacity building.

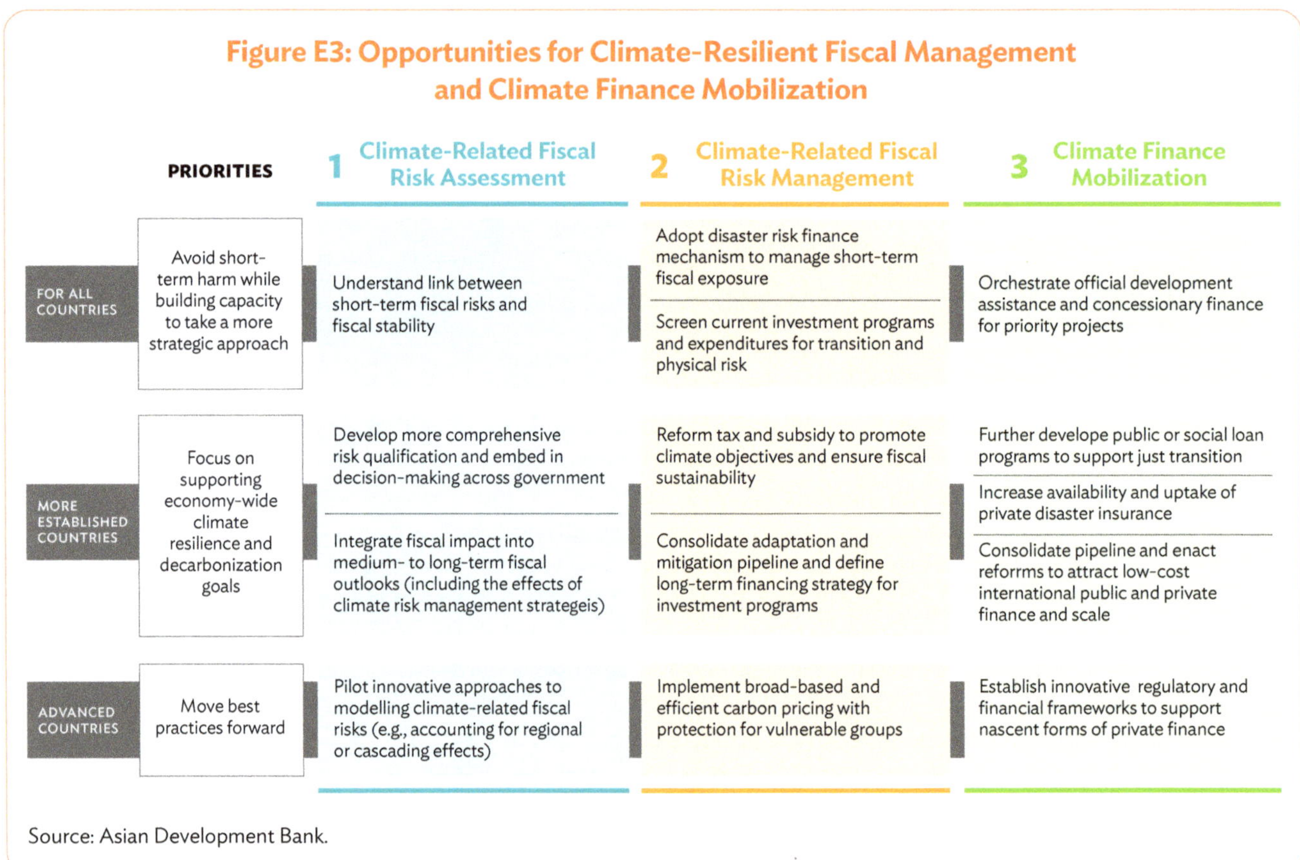

Figure E3: Opportunities for Climate-Resilient Fiscal Management and Climate Finance Mobilization

Source: Asian Development Bank.

1. Introduction

Physical risks associated with climate change are already affecting Southeast Asian countries and are set to increase rapidly. Acute physical risks, such as floods, heat waves, tropical cyclones, and droughts, already cause annual losses of $5.2 billion in the region, with extreme events causing losses as great as 10%–45% of gross domestic product (GDP) in highly exposed countries such as Cambodia, Myanmar,[1] the Philippines, Thailand, and Viet Nam.[2] There are also chronic risks, such as sea level rise, which affect Southeast Asian countries, given the concentration of population and development in low-lying plains, coastal river deltas, and coastal areas. Regardless of global efforts to reduce emissions, these risks will grow rapidly until 2030 and beyond, threatening lives, livelihoods, and broader development objectives.

The climate transition, while limiting physical risk over the long term, may also lead to significant frictions, as economies shift away from carbon-intensive products, processes, and technologies. Transition risks range from policy and regulatory risks, which can raise costs for high emissions sectors, to technology and market risks such as changing consumer trends and uptake of new technologies. Pressures on Southeast Asian countries from the transition will be substantial, with fossil fuels constituting about 80% of the region's energy mix and renewable alternatives underdeveloped.[3] Additional pressures will be felt by fossil-fuel-producing countries in Southeast Asia: for instance, the petrochemical industry accounts for 11% of Malaysia's export and the national petrochemical producer accounted for 15% of the Malaysian government's revenues between 2015–2020.[4] Southeast Asian countries are also particularly exposed to global trends in decarbonization, with high levels of export dependency.[5]

Finance and planning ministries have a critical role in responding to climate-related risks and opportunities. These ministries, responsible for government financial functions, including budget and investment management, tax and subsidy policy, debt management and financing, monitoring of public entities, and financial system regulation (Figure 1), are referred to as central finance agencies (CFAs) in this report. These core functions of CFAs are affected by trends and shocks caused by both physical and transition risks. At the same time, CFAs play a critical role in developing a fiscal system that helps accelerate the net-zero transition, thereby containing future climate change, and that promotes resilience to physical risk.[6]

---

[1]   Effective 1 February 2021, ADB placed a temporary hold on sovereign project disbursements and new contracts in Myanmar.

[2]   J. Beirne, N. Renzhi, and U. Volz. 2021. Bracing for the Typhoon: Climate Change and Sovereign Risk in Southeast Asia. *ADB Institute Working Paper Series*. No. 1223. Asian Development Bank.

[3]   D. Gielen and International Renewable Energy Agency (IRENA). 2020. *ASEAN Energy Transition Outlook*. Presentation prepared for the Singapore International Energy Week (SIEW). August 18; and I. Overland, et al. 2021. The ASEAN Climate and Energy Paradox. Energy and Climate Change. 2 (2021) 100019.

[4]   EU-ASEAN Business Council. 2023. Energy Transition in ASEAN. Singapore: EU-ASEAN Business Council.

[5]   J. Beirne, N. Renzhi, and U. Volz. 2021. Bracing for the Typhoon: Climate Change and Sovereign Risk in Southeast Asia. *ADB Institute Working Paper Series*. No. 1223. Asian Development Bank; and ASEAN. 2021. ASEAN State of Climate Change Report. Association of Southeast Asian Nations (ASEAN).

[6]   Coalition of Finance Ministers for Climate Action (CFMCA). 2023. *Strengthening the Role of Ministries of Finance in Driving Climate Action. A Framework and Guide for Ministers and Ministries of Finance.*

Figure 1: Functions of Central Finance Agencies

BUDGET AND INVESTMENT MANAGEMENT	TAX AND SUBSIDY POLICY	DEBT MANAGEMENT AND FINANCING	MONITORING OF PUBLIC ENTITIES	FINANCIAL SYSTEM REGULATION
• Provide economic and budgetary projections • Drive budgeting formulation • Executive budget • Set investment and procurement frameworks	• Design taxes, subsidies, and pricing mechanisms to raise government revenues and to incentiveize economic activities	• Develop financing strategy • Manage government debt • Develop new financing vehicles or products • Establish risk transfer mechanisms	• Monitor, manage, and regulate activities of SOEs, PPPs, SWFs, and national development banks	• Formulate and implement regulations to protect consumers, and maintain financial stability

PPP = public–private partnership, SOEs = state-owned enterprises, SWFs = sovereign wealth funds.
Source: Asian Development Bank.

Climate risks, related to both the physical impact of climate change and transition to a low-carbon economy, can affect a country's fiscal health through three main channels of impact, as illustrated in Figures 2 and 3.

- **Macroeconomic shocks.** Physical risks associated with extreme climatic events, such as drought or flooding, can lead to substantial infrastructure damage, output loss, and commodity shortages. These economic impacts may trigger increased expenditure for immediate relief needs and longer-term recovery costs, reduced tax revenues, and cause financial volatility, ultimately affecting fiscal balances and the cost of debt. For example, estimates of relief and recovery costs associated with a 1-in-200-year disaster can be as much as 20% of government expenditures (footnote 2). Similar volatility can stem from the effects of the transition—for example, the full implementation of the Carbon Border Adjustment Mechanism (CBAM) of the European Union (EU) from 2026 could conceivably lead to shocks in Southeast Asian countries' external balances, as the EU is the region's third-largest trading partner.[7]

- **Long-term macroeconomic trends.** Climate risks can also create longer-term fiscal pressures as the national and global economy transitions and adapts to physical risk. These can range from changes in productivity and growth trajectory (e.g., value of Indonesia's agriculture production could decrease by 10% by mid-century) to supply chain shifts (e.g., relocation to less at-risk geographies) and changes in resource, product, and service demand (e.g., phasing out of fossil fuels). Such shifts can affect the tax base and spending on welfare or other public services, potentially creating a need for a longer-term rebalancing to ensure fiscal positions remain sustainable (footnote 4).

- **Public and private investment.** Adaptation and mitigation programs require the deployment of capital on a very large scale across Southeast Asia: required investment in clean energy alone is projected at $150 billion per year by 2030.[8] Much of this spending will be directed toward infrastructure, requiring a mix of direct public support, incentives, and favorable regulatory conditions for private investment. Where debt is denominated in foreign currencies, this can increase vulnerability to macroeconomic shock.

7   S. Shaw. 2023. What Does the EU's Carbon Border Adjustment Mechanism Mean for Asian Economies? *The Diplomat.* 7 September.
8   A. Zarim and A. Sastry. 2024. How to Achieve a Just and Responsible Energy Transition in ASEAN. World Economic Forum Annual Meeting. 16 January.

Figure 2: Fiscal Risks Related to the Physical Impacts of Climate Change

CLIMATE IMPACT	ECONOMIC IMPACT	FISCAL IMPACT	CLIMATE RISKS TO FISCAL HEALTH

Acute short-term climate events

 **Flooding**

 **Extreme storms**

 **Drought**

 **Heatwaves**

Macroeconomic shocks

- Infrastructure damage and disruption
- Sector shocks (e.g., output loss, supply chain disruption)
- Commodity or product shortages and price volatility
- Financial sector risk

- Relief and reconstruction expenditures
- Insurance provision
- Currency for financial sector
- Liabilities for state-owned enterprises (SOEs) and public–private partnerships (PPPs)

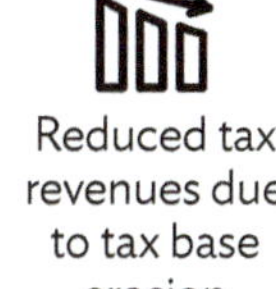

Reduced tax revenues due to tax base erosion

Chronic long-term changes in weather and climate

 **Sea level rise**

 **Chronic temperature change**

 **Ocean acidification**

 **Desertification**

Long-term macroeconomic trends

- Shift in productivity and growth trajectory (e.g., reduced agriculture yields)
- Supply chain shifts (e.g., relocation to less at-risk geographies)
- Changes in employment patterns
- Changes in income distribution

- Shift in revenue and expenditure profiles
- Shift in export/import balance
- Changes in demand for welfare payments and public services

Increased spending

Increasing public debt

Public and private investment

- Investment in resilient and low carbon technologies
- Provision of supporting infrastructure

- Direct provision of public investment (including PPPs)
- Investment by SOEs and through state-owned banks
- Incentives and regulations to support private investments, including foreign direct investment

Reduced credit rating and increased cost of debt

Note: Physical risk relates to those derived from the hazard × exposure × vulnerability framework.
In this figure climate risks to fiscal health arise from macroeconomic shocks and long-term economic trends linked to the impact of extreme weather events (acute events) and long-term climate change (chronic events).
Source: Asian Development Bank.

Figure 3: Fiscal Risks Related to the Transition to a Low-Carbon Economy

TRANSITION RISKS	ECONOMIC IMPACT	FISCAL IMPACT	CLIMATE RISKS TO FISCAL HEALTH

Policy and regulation can affect competitiveness by

- increasing cost of products through emissions pricing, regulatory changes, subsidy withdrawal etc.; and
- increasing new product and production requirements through incentives, new regulation, mandates, standards and reporting requirements thus creating pressure on downstream supply chains.

Technology risk associated with low-carbon and adaptation technology can bring disruption by

- increasing cost of transition, especially if lessons learned and economies of scale fail to deliver the anticipated cost reductions.; and
- delaying the transition, if the uptake of specific technologies hinges significantly on supporting infrastructure or downstream investment to create domestic or international demand.

Market and consumer trends can affect supply and demand or materials and products and services, as well as the financing of related projects, including

- declining demand and market price for materials, products, and services associated with carbon intensive technologies;
- increasing demand in materials, products, and services for low-carbon technologies, resulting in price increases and even supply shortages; and
- increasing reputational concerns, resulting in abrupt demand shocks or declining access to affordable financing.

Economic Impact

Macroeconomic shocks

- Sector shocks
- Commodity or product shortages and price volatility
- Financial sector instability driven by asset write-downs, rising loan repayment defaults etc.

Long-term macroeconomic trends

- Shift in sector productivity, stranded assets
- Supply chain shifts (e.g., change in demand of input factors)
- Shift in trade balance and possible increase in deficit especially for fossil fuel export-driven countries
- Changes in employment patterns
- Changes in income distribution

Public and private investment

- Investment in resilient and low-carbon technologies
- Provision of supporting infrastructure

Fiscal Impact

Macroeconomic shocks

- Compensatory payments and firm bailouts
- Currency depreciation
- Guarantees for financial sector
- Reduced credit rating and borrowing capacity

Long-term macroeconomic trends

- Shift in revenue (e.g., natural resource rents, SOE dividends) and expenditure profiles (e.g., clean energy subsidies)
- Liabilities for SOE and PPP
- Shift in trade balance and possible increase in deficit especially for fossil fuel export-driven countries
- Reduced credit rating and borrowing capacity
- Transfer payments to ensure just transition
- Demand in public job search and retraining services
- Changes in demand for welfare payments and public services

Public and private investment

- Direct provision of public investment (including PPPs)
- Investment by SOEs and SOBs
- Incentives and regulations to support private investments, including FDI

Climate Risks to Fiscal Health

Reduced tax revenues due to tax base erosion

Increased spending

Increasing public debt

Reduced credit rating and increased cost of debt

FDI = foreign direct investment, PPP = public–private partnership, SOBs = state-owned banks, SOEs = state-owned enterprises.
Source: Asian Development Bank.

To respond to these challenges, CFAs can (i) systematically assess the implications of climate risk for fiscal sustainability, (ii) adjust planning approaches and fiscal levers to manage climate-related fiscal risks and support broader adaptation and transition objectives, and (iii) mobilize public and private capital to support investment in resilient, low-carbon development. This report presents a framework for action in three steps (Figure 4):

(i) **Climate-related fiscal risk assessment:** Develop the baseline understanding of climate-related fiscal risks and embed them in fiscal planning frameworks. It includes a high-level qualitative assessment of high-impact climate risks (i.e., physical and transition), followed by a quantitative impact assessment of high-priority climate risks on fiscal functions. The risk assessment can then be embedded in decision-making on investment planning and budget management, and potentially disclosed to investors.

(ii) **Climate-related fiscal risk management:** Ensure that fiscal planning is conducive to adaptation and mitigation objectives and maintains fiscal sustainability. It includes

    (a) risk assignment to define responsibilities for managing climate risks across the economy and measures to limit and manage the CFA's liabilities;

    (b) setting investment and financing strategies to support climate resilience and decarbonization objectives;

    (c) managing revenues and spending to ensure that the climate resilience and decarbonization objectives are incentivized and the fiscal outlook is sustainable; and

    (d) adopting disaster risk finance strategies to ensure the fiscal position is robust to instability from climate risk.

(iii) **Climate finance mobilization:** Support the deployment of public and private finance for investment in adaptation and mitigation projects. Actions include interventions to increase public fiscal capacity for investment, blending public and private finance, and attracting private capital for climate goals by shaping blended finance initiatives, creating investment incentives, and fostering an enabling regulatory environment (e.g., green taxonomy, mandated climate risk disclosures, or stress testing).

This report details a framework for CFAs across Southeast Asia to address and manage climate risk. It consolidates earlier work by ADB in 2023, *Climate Resilient Fiscal Planning: A Review of Global Good Practices and Climate Resilient Fiscal Planning in Armenia,* and the Coalition of Finance Ministers for Climate Action (CFMCA) in 2023, *Strengthening the Role of Ministries of Finance in Driving Climate Action.* It establishes a framework for action and provides guidance on key priorities for CFAs across the region. The report details the framework's relevance for Southeast Asia by functions and actions, outlines key challenges, and provides examples of good practices and opportunities for climate-resilient fiscal management and climate finance mobilization. The Appendixes include case studies of good practices cited in this report along with a compendium of public resources for policymakers.

Figure 4: Framework for Action for Climate-Resilient Fiscal Management and Climate Finance Mobilization

1 CLIMATE-RELATED FISCAL RISK ASSESSMENT

1A Climate risk identification

- Identify relevant climate-related fiscal risk channels
- Assess relative materiality of risks
- Prioritize risks for quantitative impact assessment

1B Climate risk impact assessment

- Use range of techniques to quantify impact of risks on medium-term macroeconomic and fiscal trajectories and stability of public finances

1C Climate-related fiscal risk disclosure and integration into fiscal planning framework

- Publicly disclose key findings from fiscal fiscal risk assessment
- Embed climate risk assessment into decision-making processes across government

2 CLIMATE-RELATED FISCAL RISK MANAGEMENT

2A Risk assignment

- Assign risk ownership across public and private sectors to limit CFA liabilities and promote efficient risk management
- Promote or mandate risk management practices across public and private sectors

2B Climate investment and financing strategy

- Ensure investment plans are consistent with climate priorities
- Develop financing strategies
- Support the development of adaptation and transition-focused investment programs
- Embed climate consideration into public investment appraisal processes
- Embed climate consideration into public procurement processes

2C Ensuring sustainable fiscal trajectory

- Reform tax and subsidy systems to incentivize climate mitigation and adaptation programs
- Design fiscal reforms to ensure sustainable public revenue and expenditure trajectory

2D Building-up fiscal resilience against climate-related shocks

- Deploy risk retention and transfer instruments

3 CLIMATE FINANCE MOBILIZATION

3A Mobilize and deploy public funds

- Enhance capacity for public finance
- Green and expand access to public loans
- Green the investment of sovereign wealth funds

3B Mobilize and deploy private capital and international public funds

- Crowd in private captial
- Systematically green the financial sector

CFA = central finance agency.
Source: Asian Development Bank.

2. Climate-Related Fiscal Risk Assessment

This chapter reviews good practices for assessing climate-related fiscal risk. It highlights ways to integrate the climate-related fiscal risk assessment in existing fiscal management tools used to forecast revenues and expenditures. The chapter covers good practices for identifying climate risks, channels through which these impact fiscal functions, and prioritization of the risks for the subsequent quantitative assessment. It also considers approaches for quantifying the fiscal impacts of climate risks. Finally, it discusses good practices in disclosing and disseminating the insights from assessing climate-related fiscal risks. Good practices for integrating climate considerations in the budget and investment planning process are addressed in Chapter 3.

Climate Risk Identification

Context and Key Levers

A preliminary step in assessing climate-related fiscal risks is to identify potential channels through which climate risks can affect CFAs' fiscal functions. This considers how both physical and transition risks can affect economies and societies—and to what extent this creates risks for or necessitates responsive action by CFAs. While most Southeast Asian countries have a view on how physical and transition risks can shape their economies (in some cases developed in partnership with ADB or the World Bank), few currently evaluate fiscal implications systematically, considering short-, medium-, and long-term implications.[9] To develop such a view, CFAs can apply three levers:

(i)   Identify relevant climate-related risk channels to develop a comprehensive understanding of the risk exposure.

(ii) Assess the relative materiality of risks to inform a prioritization of climate risks.

(iii) Prioritize risks for quantitative impact assessment.

Key Challenges

Complexity of impact channels. A comprehensive identification of impact channels requires an understanding of an eclectic set of potential pressures from physical and transition risks on sectors of the economy, public services, and groups of people—including their effect on the level and volatility of economic activity, knock-on impacts on investment and the external balance, and implications on fiscal indicators. Many key risks will be concentrated in areas where impacts may not be fully transparent, for example in climate-related liabilities related to SOEs, state-owned banks (SOBs), and public–private partnerships (PPPs).

Complexity of relative risk ranking. In many cases, specific risks involve complex chains of exposure and vulnerability, such as how climate affects the spread of vector-borne diseases. The significance of these risks may hinge on the impact of a "tail event" (e.g., a 1-in-200-year flood) or their potential correlation with other risks (e.g., the likelihood of a state-owned miner facing distress when exports decline). Furthermore, the potential impact of any risk is subject to considerable uncertainty, depending on the future trajectory of transition and physical risks, as well as demographic, economic, and technological trends. These factors will evolve over time with rising global temperatures and declining emissions, making even qualitative comparisons of the materiality of risks challenging.

---

[9]   Examples include the Climate Risk Country Profiles of the World Bank and ADB focusing on physical risks. Accessible at https://climateknowledgeportal.worldbank.org/country-profiles.

Prioritizing risks for focus. Prioritization will inevitably require trading off the interests of different sectors, regions, and affected groups. This may involve considering equity, such as protecting vulnerable groups as part of a just transition, as well as taking into account relevant political dynamics.

Good Practices

Identify Relevant Climate-Related Fiscal Risk Channels

CFAs can leverage existing risk assessments and frameworks to ensure comprehensive risk identification. First, a starting point could be a review of existing national adaptation plans (NAPs), nationally determined contributions (NDCs), or risk assessments from technical assistance or country risk reports, including those of other countries to ensure coverage of all relevant risks and impact channels. The fiscal climate risk assessment conducted in collaboration with Armenia's Ministry of Finance and ADB can serve as a good practice reference.[10] Second, CFAs can compare the insights from the existing climate-related works with existing CFA frameworks and internationally pressure-tested frameworks spanning direct, indirect, and induced impacts of physical and transition risk on both long-term fiscal sustainability and fiscal resilience to shocks (e.g., ADB's physical climate risk framework for CFAs or the framework on climate-related transmission channels for finance ministries from the CFMCA).[11] Third, CFAs can engage sector-specific line ministries and experts to identify sector-specific impacts, and, in particular, sector-specific liabilities or expected impacts on required investments and revenues.

Assess Relative Materiality of Risks

CFAs can use high-level data sources to assess relative materiality to inform risk prioritization. CFAs can leverage readily available data from global sources to assess the relative likelihood of climate risks, and a mix of international and national data to assess relative exposure and vulnerability. This can involve collaboration with line ministries responsible for relevant areas such as climate change, natural resources, or environment, with CFAs focusing on understanding the impacts on fiscal positions.

- **Global scenario data to inform likelihood.** CFAs can leverage international global warming scenarios and associated hazard data for physical risk and transition scenarios (e.g., Network for Greening the Financial System [NGFS] scenarios, World Business Council for Sustainable Development scenarios). For physical risks, the hazard data provides insights into how the likelihood of different events will evolve over different time frames (e.g., 2030, 2040, 2050). The transition scenarios, in contrast, outline possible combinations of policy actions and associated economic outcomes, based on which CFAs can develop a hypothesis on the relative likelihood of specific economic outcomes.

- **Specific exposure and vulnerability factors.** CFAs can inform their assessment of the degree of exposure by reviewing data on specific exposure indicators (e.g., current and future sector size, sector's contribution to fiscal revenues and expenditure, revenue of exposed SOEs or PPPs) and vulnerability indicators (e.g., profitability, debt ratios, or "nature" reliance of sectors, SOEs, or PPPs, regional poverty levels, and the population's age and health indicators).

To ensure that national and local extremes are well reflected in the assessment, CFAs could explore local variations, reinforcing effects, and tail events. This can be part of a cross-government effort, potentially led by line ministries of local authorities, with CFAs focusing on fiscal consequences. Local variations can manifest in the concentration of risks (e.g., variation in severity of expected floods), exposure (e.g., densely populated areas), and vulnerability (e.g., areas with limited protective infrastructure or economically stretched sectors).

---

[10]   M. Schur, D. Manukyan, and V. Melikyan. 2023. Climate Resilient Fiscal Planning in Armenia. *ADB Sustainable Development Working Papers Series.* No. 89. Asian Development Bank.

[11]   ADB. 2023. *Climate Resilient Fiscal Planning: A Review of Global Good Practices;* and N. Dunz and S. Power. 2021. *Climate-Related Risks for Ministers of Finances: An Overview.* Coalition of Finance Ministers for Climate Action (CFMCA).

Capturing local variation will help ensure that the prioritization is just (i.e., not skewed to moderate but nationwide risks) and captures local extremes that can have significant countrywide knock-on effects. Reinforcing effects are not always quantified in climate risk models and can thereby at least be reviewed qualitatively. An example of a reinforcing effect is the higher risk of flooding under deforestation or after prolonged heatwaves. Lastly, assessing the potential impact of tail events provides a view on the maximum possible impact and the potential cost of inaction.

Prioritize Risks for Quantitative Impact Assessment

To ensure a high-quality prioritization process, the prioritization mechanism can be clearly defined in advance and be applied consistently across all physical and transition risks. Elements of the prioritization mechanisms, such as methods to aggregate relative likelihood, exposure, and vulnerability, or weights given to specific vulnerabilities as well as future exposures, can be defined and aligned in advance. The development of the prioritization mechanism and the review of the outcomes can be a consultative process, involving relevant stakeholders (e.g., other line ministries, authorities involved in NAP, NDC, long-term climate strategies (LTSs), and just transition framework development) to ensure alignment with previous work.

Climate Risk Impact Assessment

Context and Key Levers

To understand the magnitude of the risk exposure, CFAs can use a range of techniques to quantify the impact of climate-related risks on medium-term macroeconomic and fiscal trajectories and on the stability of public finances. Some Southeast Asian countries have assessed the impact of climate change on specific economic positions as part of their NAP, NDC, and LTS development. For instance, Viet Nam identified in its NAP that the production of rice could decrease by 8.8% and maize by 18.7% in 2030.[12] Indonesia found that the low-carbon transition will require investments of $4.0 billion to $9.8 billion per year in 2025–2045, potentially raising GDP growth by 1%–2% a year relative to no transition.[13] These projections account for the effect of physical risk on the productivity of sectors such as agriculture and infrastructure and make some allowances for the related costs of adaptation in sectoral investment plans. However, there remains a need to fully "climate-proof" the national plans to ensure their benefits are realized. More broadly, while understanding the magnitude of impacts on economic variables such as employment and investment is an important first step, further work is required to understand the fiscal implications for CFAs to effectively prioritize actions.[14]

Key Challenge

Limited data availability and technical challenges in economic modeling of climate risks. Quantitative modeling in this area can be highly technically challenging with limited data available in certain key areas. Modeling direct damages from physical hazards can be highly complex, requiring the assessment of damages of events of varying severity, considering local exposure and vulnerability, and aggregating the outcomes into summary statistics (e.g., average annual damages and exceedance probability curve). To analyze other economic effects, a host of further impacts can be assessed, including impacts on health and infrastructure systems, business interruptions, losses propagating along supply chains, and long-term impacts on the cost of capital and investment. These, in turn, are translated into fiscal impacts, accounting for factors such as disaster relief and reconstruction spending, tax revenue losses, and automatic stabilizers in welfare systems. Such exercises are not only technically challenging, but may also

---

[12]   Government of Viet Nam, Ministry of Natural Resources and Environment. 2022. *National Adaptation Plan for the Period 2021-30, with a Vision to 2050*.

[13]   Government of Indonesia, Ministry of National Development and Planning (Bappenas). 2019. *Low Carbon Development: A Paradigm Shift Towards a Green Economy in Indonesia*.

[14]   GIZ. 2021. *Macroeconomic Models for Climate Resilience*.

require climate, economic, or fiscal data that is unavailable or of low quality. Similar complexities could arise when modeling transition impacts such as the systematic consequences of carbon pricing and the complex dynamics of disorderly transition (e.g., higher carbon prices increasing corporate default risks and disproportionately burdening lower-income households). Some CFAs may lack the capacity and resources to develop such assessments.

Good Practices

Use a range of techniques to quantify the impact of risks on medium-term macroeconomic and fiscal trajectories and the stability of public finances

CFAs can consider three key dimensions of good practice when developing the climate-related fiscal risk assessment: scope and coverage, sensitivity and stress testing, and forward-looking implications. A good fiscal risk assessment goes beyond explicit and implicit liabilities by also capturing second-order impacts on the economy. In the best case, CFAs deploy macroeconomic models that link first-order impacts (e.g., price increase due to carbon pricing, physical shocks) to economic knock-on effects across other sectors. The Global Best Practice report provides an overview of commonly used macroeconomic models.[15] Alternatively, Southeast Asian countries can deploy approaches focused on event-specific impact assessments. For instance, post-disaster needs assessment (PDNA) can help quantify the damages (first order impact) and losses due to disrupted access to goods and services (second order impact) based on real post-disaster data instead of estimated losses. Such an assessment can inform the potential impact of future disaster events of a similar magnitude or even a broader stress test. Generally, CFAs could consider modeling methods that are transparent in their assumptions and easy to explain to ensure the confidence of decision-makers in their results (footnote 6).

Ultimately, a macroeconomic model can establish a clear link between the resulting government liabilities, the second-order economic impacts, and strategic fiscal health indicators. Tying the analysis to fiscal health outcomes will allow CFAs to assess the aggregate fiscal implications based on a clear set of indicators that can be monitored over time. CFAs can select fiscal health indicators that align with the government's strategic priorities and existing fiscal planning processes. The set of fiscal health indicators can include **single or composite indicators**, which are revenue and expenditures (both linked to implicit and explicit liabilities) or the debt–to–GDP ratio; and **short- or long-term indicators**, which are gross external financing requirements or the debt–to–GDP ratio.

To assess fiscal resilience, CFAs can use scenario modeling or stress tests based on well-recognized future assumptions and conditions. Scenario modeling and stress tests are well-established approaches that can help assess the resilience of the fiscal system under different future "states." CFAs can develop their country-specific scenarios, drawing on national disaster data or assumptions from internationally well-established and pressure-tested scenarios (e.g., Network of Greening the Financial System [NGFS] transition scenarios, the Representative Concentration Pathway [RCP] and associated global warming scenarios of the Intergovernmental Panel on Climate Change [IPCC]). CFAs can codevelop stress tests and country-specific scenarios in collaboration with the national monetary and financial supervisory authorities. For example, the Monetary Authority of Singapore (MAS) defined and included climate-related scenarios in its 2022 Industry-Wide Stress Test exercise for the financial sectors.

CFAs can review the effect of asymmetrical shocks on the economy (e.g., a climate-related shock to the agriculture sector) to pressure-test fiscal forecasts by forecasting tax revenue at a sector, rather than an aggregate level. CFAs can opt for such a sector-specific approach if model or data limitations do not allow a "full integration."

---

[15]   For example, dynamic input-output models, computable equilibrium models, computable general equilibrium models, integrated assessment models, dynamic stochastic general equilibrium models, etc.

CFAs can use forward-looking approaches to capture the future evolution of impacts. Both physical and transition risks are likely going to increase in the mid to long term. Through forward-looking analysis covering various time horizons (e.g., 2030, 2050), CFAs can develop a quantitative understanding of the implications on the economic structure, the fiscal system, and the need for mitigation and adaptation project investments. The future-based assessment will be based on commonly used socioeconomic scenario assumptions, e.g., population growth, and urbanization rates, and climate policy assumptions.

CFAs can refresh climate-related fiscal risk assessments annually using updated assumptions from fiscal forecasts that are typically made annually. A deeper overhaul of the risk assessment, updating climate trajectories and the relationship between fiscal outcomes and physical and transition risk, would only be necessary every 3 to 5 years, given more gradual updates in the relevant evidence base: for example, IPCC assessments are done every 5 to 7 years.

The following are five recent examples that implemented aspects of good practice. They include both physical and transition risks and a variety of data sources and modeling techniques, ranging from the analysis of historical data to risk simulation to long-term macroeconomic scenario modeling:

- The Lao People's Democratic Republic (the Lao PDR) conducted a PDNA across 11 sectors to quantify the damages and losses after the historic flash floods in 2018.[16] The PDNA also provided insights on the impact on the Lao PDR's fiscal health. It found that the budget deficit increased by 0.5 percentage points in 2018 and that 1.5% of government expenditures are expected to be redirected for medium- to long-term recovery. The results could be leveraged for future fiscal stress tests (Box A1).

- For preparing its LTS, the Government of Indonesia used the Asia-Pacific Integrated Model/Computable General Equilibrium model to analyze the impact of mitigation in the agriculture, forestry, and other land uses; waste; industry; and energy sectors. Beyond economic growth, the assessment estimated the expected increase in government expenditures, national investment levels (not differentiating private and public investments), and impact on employment rate (which could be linked to expected tax revenue). The assessment covered three mitigation scenarios and estimated the impacts until 2050 (Box A2).[17]

- The Government of Indonesia assessed the potential impacts of the early retirement of existing coal assets on the national fiscal position, including revenue impact from reduced tax and increased tax from other sectors and spending impact from increased compensation and social benefits (Box A3).

- The Task Force on Climate Development and the International Monetary Fund (IMF) developed a parametric macroeconomic model to analyze the effects of carbon pricing in the People's Republic of China (the PRC) on the Indonesian economy and Indonesia's balance of payment across three NGFS scenarios until 2050 (Box A4).[18]

- The IMF deployed an econometric model to estimate the expected fiscal impacts of physical climate change on Armenia's GDP, employment, and debt–to–GDP ratio across three climate scenarios until 2070 (Box A5).[19]

---

[16]   Government of the Lao PDR. 2018. *Post-Disaster Needs Assessment, 2018 Floods, Lao PDR.* It covered social sectors (e.g., housing and settlements, education), productive sectors (e.g., agriculture, industry, tourism), and infrastructure sectors (e.g., transport, electricity).

[17]   Government of Indonesia. 2021. *Indonesia Long-Term Strategy for Low Carbon and Climate Resilience 2050.* UNFCCC.

[18]   R. Gourdel, I. Monasterolo, and K. P. Gall. 2022. *Climate Transition Spillovers and Sovereign Risk: Evidence from Indonesia.*

[19]   J. Harris, et al. 2022. Armenia: Quantifying Fiscal Risks from Climate Change. *IMF Country Report No. 22/329.* International Monetary Fund.

Climate-Related Fiscal Risk Disclosure and Integration into Fiscal Planning Frameworks

Context and Key Levers

Public disclosure of climate-related fiscal risks can enhance cross-government responses and improve investor confidence in countries' management approaches. Public disclosure can enhance government coordination and collaboration on climate risk management by supporting a shared understanding of its fiscal impacts (footnote 16). It can also help boost confidence in the government's ability to manage the climate risks, especially in the Asia and Pacific region where climate risk has been projected to lower the credit rating of 63 sovereigns by 2030.[20] To further improve climate-related fiscal risk management, CFAs can integrate the climate risk assessment outcomes into the fiscal planning frameworks. Integrating the climate risk assessment into fiscal planning frameworks can help ensure that the budget forecast and allocation of resources is cognizant of potential revenue shortfalls and contingent expenditures (footnote 16). Chapter 3 highlights critical decisions taken on the base of climate-related fiscal risk assessments in investment and finance strategies, revenue and spending management, and disaster risk finance.

Key Challenges

Alignment across government. External stakeholders use fiscal disclosures to assess countries' risk exposure and management. CFAs will require alignment across bodies charged with managing key risks in advance of making any disclosure.

Diversity of audience. CFAs have a broad audience, each requiring different types and formats of risk information. Investors are more likely to require a succinct report of all fiscal risks, including climate risk, and a synthesis of the potential magnitude of risks. In contrast, line ministries, subnational governments, and other government agencies might require raw data, tools, and other guidance to incorporate climate risk assessment insights into their analyses and decision-making processes. This may involve feeding fiscal risk information into specific platforms and decision tools used by government agencies.

Lack of risk management strategy. Public disclosure of climate-related fiscal risks can be accompanied by announcements or publications of a convincing and robust risk management strategy and the commitment to implement it. A lack of transparency on the risk management strategy while disclosing fiscal risks may undermine confidence and trust in governments, public institutions, and financial markets, potentially resulting in higher financing costs.

Capacity constraints for integrating climate risks into fiscal planning. The successful integration of fiscal risk assessments largely depends on the sophistication of existing economic planning frameworks and the consistent availability of qualified staff. These staff members are essential for driving integration and maintaining the evidence base through annual updates and periodic revisions to the methodology and data. Moreover, institutional, capacity, and data gaps in existing systems could further prevent the effective integration of climate goals into fiscal management processes. For example, in Cambodia, the public investment management (PIM) function is in its early stage of development.[21] In Indonesia, the IMF's Public Investment Management Assessment (PIMA) in 2019 highlighted the need for improved coordination among public investment institutions (i.e., ministries and local governments).

---

[20]   P. Klusak, et al.. 2021. *Rising Temperatures, Falling Ratings: The Effect of Climate Change on Sovereign Creditworthiness.* Bennett Institute.

[21]   Public Expenditure and Financial Accountability. 2021. *Cambodia: PEFA Performance Assessment Report 2021.*

Good Practices

Publicly Disclose Key Findings from Fiscal Risk Assessment

CFAs can publicly disclose the climate-related fiscal risk alongside other sources of fiscal risk. To restrict the downsides of public disclosure of fiscal risks, risk management strategies for both climate-related fiscal risks and other fiscal risks can be integrated into the reporting of risks. For instance, the Government of the Philippines reports climate-related fiscal risks in its annual fiscal risk statement (Box A6). The Philippines's Fiscal Risk Statement 2023, prepared by the Development Budget Coordination Committee with representation from the Department of Finance and the Department of Budget and Management, provides insights on, for example, damages associated with past disasters triggered by natural hazards and the resulting government expenditures, the required investment need for the recovery and rehabilitation of affected regions until 2025, the impact of chronic climate events on agriculture productivity (e.g., approximately ₱8 billion in reported damages due to El Niño) and the strategy and actions of the government to reduce and manage climate-related fiscal risks. CFAs can additionally include climate-related fiscal risks in communication papers shared across the government during the budget preparation process (e.g., budget framework papers or the opinion of CFAs on the fiscal strategy and risks) (footnote 6). To ensure coherence, a good practice is to also include fiscal risk insights in NAP and NDC disclosures.

To ensure well-coordinated and targeted communication, CFAs can create a central source for fiscal risk information and develop use-case-specific dissemination approaches. A good practice dissemination approach could include the following:

- **For the public (including citizens, private sector, and investors):** Climate risk disclosure as part of the fiscal risk statement, covering key dimensions outlined by the Task Force on Climate-related Financial Disclosures (TCFD) or the International Sustainability Standards Board recommendations (e.g., risk metrics and targets for mitigation and adaptation projects, strategy to reduce and manage risks, risk management, and governance processes).

- **For line ministries and subnational government:** Raw data on and the visualization of regional and sector-specific hazards, disaggregated data on economic impact, access to models and tools to assess impact, regional or sector-specific revenues, and expenditures under varying policy scenarios.

To maintain high-quality service, CFAs can consider establishing a team dedicated to managing the relevant data and models and providing support for users.

Embed Climate Risk Assessment into Decision-Making Processes Across Governments

The way in which CFAs integrate climate-related risks into the fiscal planning process will vary country by country, but there are several good practices that CFAs can consider:

- **Build on existing planning processes.** The climate risk assessment could ideally be reflected in key macrofiscal frameworks, such as the medium-term fiscal framework (MTFF) and the medium-term budget framework (MTBF). The underlying modeling approach to develop the MTFF can vary.[22] Broadly speaking, CFAs can integrate climate risk parameters into existing macrostructural models or adjust the GDP growth forecast by the expected climate risk impacts to update forecasts in revenues and expenditures.[23]

---

[22]   ADB. 2024. *A Governance Framework for Climate-Relevant Public Investment Management.*
[23]   IMF. 2020. *Challenges in Forecasting Tax Revenue.*

- **Enable integration at subnational or sector level.** To best inform decision-making across the government, the CFAs can help line ministries or subnational governments to integrate climate risk insights in sector-specific or regional planning frameworks.

- **Align update frequency with the budget process.** To integrate climate risk into fiscal planning, the climate risk assessment cannot be a one-off exercise but should be replicable at least once per year. Ideally, CFAs can consider the replication feasibility when defining the assessment methodology.

- **Issue direction on the use of risk assessment.** Chapter 3 considers how fiscal risk information can be used by decision-makers within CFAs, which can be governed by the processes of the CFAs. For users outside CFAs, including line ministries and local authorities, the adoption of such processes can be facilitated by the CFA issuing specific guidance or, potentially, mandates on circumstances in which the risk assessment can be used.

In contexts of wider requirements to strengthen public financial management (PFM) systems, CFAs can consider realistic ways to integrate climate risk assessments within them. Within Southeast Asia, countries exhibit varying degrees of maturity in their PFM systems and there are opportunities for enhancement, particularly in relation to cross-government coordination, stakeholder participation, consistency, transparency, accountability, and independent oversight (footnote 23). In the case of investment planning (Chapter 3), the IMF has developed the Climate-Public Investment Management Assessment (C-PIMA) framework to help countries improve institutions and processes for PIM in relation to climate-resilient infrastructure.[24] To avoid creating additional burdens, CFAs could start with modest steps, such as implementing a simple set of screening questions to prevent investments that might result in "lock-in" emissions and exposure (footnote 6).

---

[24]   PEFA. 2020. *Climate Responsive Public Financial Management Framework (PEFA Climate).*

3. Climate-Related Fiscal Risk Management

This chapter reviews good practices for managing fiscal risk and developing supporting fiscal strategy. It focuses on good practices for assigning climate risk management across the government as well as private- and state-owned entities with the aim to reduce liabilities. It also considers good practices for integrating climate considerations in public investment strategies; the development of investment pipelines; and public budget, investment, and procurement programs. This chapter sets out good practices for managing revenue and expenditure to appropriately incentivize mitigation and adaptation projects, while ensuring the fiscal trajectory is sustainable. Finally it concludes by reviewing good practices to ensure the fiscal position is resilient to climate-related shocks.

Risk Assignment

Context and Key Levers

Explicit and implicit climate-related risks represent a significant threat to the fiscal positions of many Southeast Asian countries. Explicit liabilities are established by the law of contract, including direct liabilities that apply in any event, such as grants and subsidies by budget law, or contingent liabilities that apply only if a specific event occurs, such as a state guarantee for borrowings at an oil and gas SOE that can no longer repay the debt. Implicit liabilities, on the other hand, arise from perceived obligations of the government to support certain groups or actions and also include direct liabilities, such as costs of future public investment projects on flood protection, and contingent liabilities, such as disaster relief.[25] A recent study revealed that greater vulnerability to climate change has a sizeable effect on sovereign bond yields, suggesting that Southeast Asian countries are already paying a climate risk premium. As climate change progresses, countries such as Indonesia, the Philippines, Thailand, and Viet Nam could face higher borrowing costs and greater constraints in financing investments related to climate resilience (footnote 2). This could exacerbate already high financing costs in emerging and developing countries that are at least twice as high as in advanced economies.[26]

Risk assignment is a foundational step in managing these risks, by limiting CFAs' exposure and incentivizing risk management by others. By explicitly defining and distributing risk management responsibilities, CFAs ensure that each risk owner, whether it is subnational governments, SOEs, corporates, or households, is aware of its obligations in mitigating climate-related risks. This structured approach not only helps limit implicit liabilities with regard to the CFAs, but also fosters a culture of accountability and preparedness. For example, if firms are responsible for bearing the costs of weather-related damages and losses, they will have incentives to invest in resilience measures (e.g., asset-level protection, resilient supply chain) or purchase insurance. This will reduce CFAs' contingent implicit liabilities and reduce the vulnerability of different sectors with regard to climate-related disruptions.

There are two key levers for CFAs to implement effective risk assignment:

(i) **Assign risk ownership.** Across the public sector, CFAs can delineate responsibilities for climate-related risk management among line ministries, departments, subnational governments, and SOEs. For the private sector, CFAs can set out recovery spending principles and strategies to prevent moral hazards and reliance on government bailouts.

---

[25] H. Brixi. 1989. Contingent Government Liabilities: A Hidden Risk to Fiscal Stability. *World Bank Policy Research Working Paper.* No. WPS1989. World Bank Group.

[26] International Energy Agency (IEA). 2024. *World Energy Investment 2024.*

(ii) **Promote or mandate risk management practices.** Once risk ownership is defined, CFAs can utilize a range of instruments to limit and manage liabilities. These include: "direct" controls, ceilings, or caps that could be used to limit CFAs' total exposure (e.g., a cap on guarantees to limit explicit liabilities to SOEs); "indirect" measures, such as regulations and incentives, to mandate or encourage risk-reducing behavior of risk owners; and the use of risk transfer mechanisms, such as insurance on public assets to limit the need for recourse for public funds.[27]

Key Challenge

Capacity constraints to manage climate risks. In practice, it can be difficult for CFAs to credibly commit to the arrangement of climate-related fiscal risks assigned and managed by other governmental bodies and private- and state-owned enterprises, given pressure to provide support ex-post. Such pressure could, for example, manifest where a subnational government that experienced a disaster triggered by natural hazards requests a bailout to support essential services, or an SOE threatened by the transition requests an equity injection to protect jobs. Furthermore, stakeholders may struggle to effectively manage risks that have been assigned to them. Subnational governments may lack the capacity to understand and respond to physical risk information, which can be highly technical, while firms and employees in carbon-intensive industries may lack access to the technologies or skills required to make a successful transition. A particular challenge relevant to Southeast Asian countries is access to affordable insurance, with many countries facing high Climate Risk Index analyses (e.g., the Lao PDR, the Philippines, Thailand, and Viet Nam).[28] A study in Viet Nam on households' willingness and ability to purchase flood microinsurance highlighted that a significant share of households was unable to afford such insurance.[29]

Good Practices

Assign Risk Ownership

A first principle in assigning risk ownership is to align it with management capacity. For example, the risk layering approach in the Philippines categorizes risks into three levels and assigns risk owners accordingly, with high-frequency, low-severity risks covered at national and local levels while mid-frequency, mid-severity risks are addressed by contingent financing from development partners (Box A7). CFAs can also provide resources to support risk owners in building the necessary risk management capacity. For example, in Indonesia, climate budget tagging tracks the portion of government spending allocated to climate change actions: this can be used to determine access to resources for risk owners to fund mitigation or adaptation activities.[30]

To enhance the credibility of risk assignment, detailed recovery mechanisms for stress scenarios can be established in advance, potentially involving incentives for earlier risk reduction efforts. More granular information on what will and will not be covered by CFAs in the event of a severe loss can bolster the credibility of a risk assignment, thus encouraging risk owners to assume responsibility for reducing and managing risks. This can be enhanced by specific incentives baked into approaches to relief or recovery. For example, the Laoag City Disaster Risk Reduction Management Plan office in the Philippines provided funding to purchase drought-resistant variety crop seeds to help farmers mitigate potential losses from droughts.[31] Taking this a step further and linking potential relief payouts with by adopting adaptation measures would further enhance incentives for risk management.

---

[27]   C. Towe et al. 2016. *Analyzing and Managing Fiscal Risks – Best Practices. IMF Policy Paper Series.* International Monetary Fund.

[28]   Germanwatch. 2021. *Who Suffers Most from Extreme Weather Events? Weather-Related Loss Events in 2019 and 2000-2019.*

[29]   R. Brouwer, et al. Modeling demand for catastrophic flood risk insurance in coastal zones in Vietnam using choice. *Environment and Development Economics.* 19 (2). pp. 228-249.

[30]   D. Rulliadi. 2019. *Climate Budget Tagging and Green Sukuk / Islamic Bond: An Indonesian Experience.* Presentation at the Sherpa's Meeting, Coalition of Finance Ministers for Climate Action. 21-22 February.

[31]   E. J. Guillermo. 2024. *Agri Office in Laoag City Encourages Farmers to Use Drought-Resistant Crops. Philippine Information Agency.* 31 January.

Although risks can be effectively assigned to different entities, a central oversight body can also monitor overall risk levels and understand interactions between different risks. This role could be delegated to either a dedicated unit or a high-level oversight committee, with a mandate to establish climate-related fiscal risk indicators (Chapter 2), monitor risk evolvement, and evaluate the adequacy of existing risk management practices (footnote 29).

Promote or Mandate Risk Management Practices

To encourage risk owners to manage risks cost-effectively, CFAs can promote a balanced set of requirements, including insurance, risk reduction, and risk retention actions. An approach based entirely on insurance is likely to be expensive and ineffective for risk owners, as compared to a mix of risk transfer, investment in protection, and fiscal buffers. While there is no one-size-fits-all approach to balancing these actions, CFAs can develop a fit-for-purpose strategy by considering three main criteria: evidence base, cost-effectiveness, and adaptability (footnote 9). To the extent risk transfer is pursued by risk owners, CFAs can encourage the uptake of a variety of measures (e.g., conventional indemnity insurance policy, parametric insurance, and catastrophe [cat] bonds) at both the local and central levels to meet overall needs in terms of speed of payouts, coverage, flexibility, and cost. For example, in the Philippines, risk layering with clear distribution rules combines local and central risk transfer and risk retention (Box A7).

For SOEs specifically, a good practice approach can involve not only mitigating risks but also proactively identifying and leveraging new opportunities. While many SOEs in Southeast Asia are major sources of emissions and thus are vulnerable to transition risks, they are also significant investors, especially in agriculture and infrastructure, which can support decarbonization and resilience objectives.[32] The expertise, skills, and capacity of SOEs can also be leveraged to support expansion into strategic green energy sectors, as seen in the case of Petrovietnam and its subsidiary Petrovietnam Technical Services Corporation's expansion into offshore wind (Box A8). The main steps and entry points for countries to adopt climate action policies in relation to SOEs include the following (footnote 34):

- conducting a climate-informed state ownership policy assessment to ensure alignment with a country's national climate change commitments;

- conducting a climate diagnostic of the SOE sector portfolio to identify gaps and areas that need reform;

- formulating a climate strategy and implementation plan for the SOE sector; and

- determining monitoring and evaluation, and audit and disclosure requirements, including for regular climate assessments and diagnostic reviews.

Climate Investment and Financing Strategy

Context and Key Levers

In many countries, there remains a disconnect between sector strategies and financing plans. Most Southeast Asian countries have developed national strategies and sector plans to address climate change, such as the Lao PDR's National Strategy on Climate Change, the NAP of Thailand, and the National Climate Change Strategy of Viet Nam (footnote 5). However, there is limited linkage between these climate plans and broader investment programs, including how investment will be supported through public and private capital. Furthermore, to avoid locking in emissions and exposure to climate risks, there is a need for wider spending, and investment decisions can be systematically screened for both physical and transition risks.

---

[32] A. I. De Kleine Feige. 2021. State-Owned Enterprises and Climate Action. *World Bank Working Paper* No. 164952. World Bank Group.

CFAs can help shape climate investment and develop the associated financing strategies through the following five main levers. It is critical that investment and financing plans are informed by the climate-related fiscal risk assessment, especially on the implications for sovereign debt trajectories:

(i) **Ensure investment plans are consistent with climate priorities.** CFAs could play a coordinating role in aligning different sector investment plans with climate priorities, ensuring they account for fiscal-related impacts and collectively form a mutually synergistic, long-term program of investment.

(ii) **Develop the financing strategies.** CFAs could develop clear financing strategies for climate-related investment plans and assessing resource requirements, determining public- vs private-led split and appropriate investment models (e.g., PPP).

(iii) **Support the development of adaptation and transition-focused investment programs.** CFAs could fund project preparation and capacity building within line ministries.

(iv) **Embed climate considerations into public investment appraisal processes.** CFAs could integrate climate considerations into annual budgets and medium-term expenditure frameworks. Key entry points for CFAs in the budget process include the budget framework paper, updates to the MTFF and MTBF, line ministry guidance and submissions, budget circulars, and review of final submissions (footnote 6). Key entry points in the public investment process include integrating physical risk analysis and management into the evaluation of infrastructure projects, and incorporating resilience and emission criteria into investment appraisal.[33]

(v) **Embed climate consideration into public procurement processes.** CFAs could establish policy frameworks for green public procurement (GPP), introduce environmental standards in procurement selection and award criteria, and assist procuring entities in incorporating climate objectives into procurement procedures (footnote 6).

Key Challenges

Trade-offs in balancing short-term development goals with long-term transition and climate adaptation programs. Given limited public resources, this balancing act requires careful consideration of the trade-offs between different sectors and regions—and may involve real or perceived conflicts between climate action and immediate-term economic development (e.g., where jobs depend on unsustainable practices in agriculture).

Uncertainties complicating the development of investment plans and the associated financing strategies. Uncertainty about the transition pathway, future socioeconomic trajectories, evolving technologies, and climate solutions make it challenging to develop robust and forward-looking investment plans. This uncertainty is exacerbated by a lack of reliable data and methodologies to estimate costs of climate actions accurately. This is especially the case for adaptation solutions, where investments are site-specific and will vary depending on the framing and objectives set.[34]

For the CFA to play a coordinating role, institutional reforms or capacity development may be required. The absence of a clear mandate for CFAs or a lack of coordination between various institutions may impede the consistent application of climate considerations across sectors and projects.

---

[33]   R. Delgado and H. Eguino, eds. 2023. *Fiscal Policy for Resilience and Decarbonization: Contributions to the Policy Dialogue.* *Technical Note.* No. IDB-TN-2652. Inter-American Development Bank.

[34]   United Nations Framework Convention on Climate Change (UNFCCC). 2022. *Efforts of developing countries in assessing and meeting the costs of adaptation: Lessons learned and good practices.*

Good Practices

Ensure Investment Plans Are Consistent with Climate Priorities

CFAs can actively coordinate across central governments and with subnational governments to ensure that climate priorities are consistently reflected across sector and subnational investment plans. Public investment processes involve numerous stakeholders, including national (Ministry of Finance, Ministry of Planning and Investment, line ministries), regional, and local governments, SOEs, private companies, and PPP entities.[35] As CFAs will have a central vision for various investment programs, they can coordinate and facilitate the effective integration of climate priorities into the PIM. CFAs can also provide the regulatory and oversight framework for SOEs to ensure that their climate-related investments are consistent with national climate policies and guidelines.

In particular, the early involvement of CFAs in the planning process, which sets clear standards, is key to developing investment programs consistent with climate priorities. The IMF's C-PIMA lists "planning" as the one priority area for creating climate-resilient and low-carbon infrastructure (Box 1). Sector plans and the associated investment portfolios can be aligned with climate objectives and the planning phase is especially critical for integrating climate considerations into spatial planning and construction requirements (footnote 37). Early involvement in the planning process could allow CFAs to anchor a holistic approach across different sectors and help CFAs develop a medium- to long-term programmatic vision. For adaptation, this would appropriately sequence and balance measures that focus on pre-disaster mitigation and prevention, preparedness, response, rehabilitation, and reconstruction, with the CFA potentially coordinating plans from a range of line ministries and local government.

Box 1: IMF's Climate-Public Investment Management Assessment

The International Monetary Fund (IMF) has developed the Climate-Public Investment Management Assessment (C-PIMA) to help countries improve institutions and processes for public investment management in infrastructure. It is designed around the following five priority areas that are seen as key for creating climate-resilient and low-carbon infrastructure:

(i) **Planning.** Aligning national and sector plans with climate objectives; this phase is key for integrating climate considerations into spatial planning and construction requirements.

(ii) **Coordination.** Coordination across government layers, state-owned enterprises, public–private partnership entities, and ensuring collaboration between the public and the private sectors.

(iii) **Appraisal and selection.** Integrating climate-related analysis of mitigation and adaptation impacts of investments to support decision-making.

(iv) **Budget and portfolio management.** Budgeting for and reporting on green investment and maintenance allocations through the annual budget and other fiscal instruments such as the medium-term expenditure framework and the government's financial statement; incorporating climate objectives into asset management and ex-post audit and review.

(v) **Fiscal risk management.** Incorporating climate-change-related risks into fiscal risk analyses and disaster management strategies.

Source: P. Mauro et al. 2021. Strengthening Infrastructure Governance for Climate-Responsive Public Investment. *IMF Policy Paper* No. 2021/076. International Monetary Fund.

---

[35]   G. Schwartz. 2023. Green Public Investment Management: Governance Aspects of Strengthening Infrastructure Sustainability. *The Governance Brief.* No. 50. Asian Development Bank.

Develop Financing Strategies

CFAs can match sources of finance with sector risk characteristics. The High-Level Advisory Group on Sustainable and Inclusive Growth categorizes investments into four types requiring different financing sources: (i) bankable investments, which can be financed by the private sector, mobilized by incentivizing policies and regulatory frameworks (Chapter 4); (ii) riskier investments, which are attractive to private investors but require de-risking, technical assistance, or grants (Chapter 4); (iii) public good investments, which include some investments in adaptation projects and are not commercially viable without concessional or grant support; and (iv) social investments, which, to support a just transition, can be financed by public sources.[36] This approach is demonstrated in Cambodia's Long-Term Strategy for Carbon Neutrality, which outlines investment themes for private investment and estimates a total private investment need of $1.4 billion per year by 2050 (Box A9).

To ensure the feasibility of the financing strategy, CFAs can develop supportive regulatory policies. One effective strategy is to reform the PPP framework. Recognizing the crucial role of the private sector in supporting transition, the Department of Finance in the Philippines has enhanced its PPP framework to encourage investment into previously restricted sectors, such as renewable energy.[37] These supporting regulatory reforms are essential to facilitate the financing of diverse investment programs, especially when aiming to leverage private sector participation.

Once the strategy is developed, clear public communication about climate-related investment needs, risk analysis, and mitigation strategy is critical for providing certainty to both investors and companies. This transparency helps private sector players understand opportunities, fostering a conducive environment for investment, and can be combined with public disclosures of climate risk assessments (Chapter 2).

Support the Development of Adaptation and Transition-Focused Investment Programs

CFAs can bring in cross-sector insights on project development challenges and bankability barriers to support targeted project preparation and capability building. CFAs could start by helping various ministries and other government agencies identify potential projects and make them bankable. Given their overarching perspective and coordinating role, CFAs are well-positioned to identify common hurdles preventing the development of adaptation and transition-focused investment programs and to support the sourcing of technical assistance to overcome these hurdles. They can also play a crucial role in linking strategic national projects with specific project preparation facilities offered by multilateral development banks (MDBs) and international financial institutions (IFIs).

Embed Climate Consideration into Public Investment Appraisal Processes

CFAs could leverage the initial strategic phase of the budget formulation process to provide clear guidance and prevent the late-stage "blocking" of resource allocations (footnote 6). By setting clear priorities and ensuring alignment with national climate plans early on, the budgeting process can effectively support allocation toward climate priorities. This is demonstrated in the Philippines' climate budgeting reform, where climate considerations are integrated into the initial process through a series of reporting and review loops (Box A10).

Similarly, for public investment, clear guidelines for embedding climate considerations at the project identification and design stages are essential to prevent late-stage design reversals. The Philippines serves as a good practice example in this area, having integrated climate considerations into the assessments required to develop PPP projects (Box A11). To facilitate rapid rollout and ensure that climate is adequately considered without overburdening the process, the level of assessment could be adjusted to project size—for example, simple screening questions for smaller projects and detailed appraisals for large strategic projects.[38]

---

36   World Bank Group. 2023. *The Big Push for Transformation through Climate and Development: Recommendations of the High-Level Advisory Group on Sustainable and Inclusive Recovery and Growth.*

37   Government of the Philippines, Department of Finance. 2023. Diokno: PH's Nationally Determined Contribution and National Adaptation Plan to Resolve Info Gap with Climate Finance Providers. 2 December.

38   World Bank. 2022. *Reference Guide for Climate-Smart Public Investment.*

These guidelines can include specific technical guidance on approaches to investment appraisal. Many CFAs set standards for public investment appraisals across the public sector, which can be expanded to include specific considerations related to adaptation or mitigation projects. Good practice case studies related to this are set out in the Appendixes. This guidance covers topics including the use of scenarios, flexible pathways of adaptation project investment, and risk assessments, aspects of which can be calibrated to be consistent with macrofiscal climate risk assessments.

Establishing clear public accounting standards and unified criteria for budget classification is vital for facilitating adoption, monitoring, and reporting. These standards could also enable the implementation of an output-based budget framework with appropriate performance indicators that enable the measurement and monitoring of climate-related spending outcomes, informing the selection of budget proposals and ensuring that the "green" budget generates real impact.[39]

Embed Climate Consideration into Public Procurement Processes

The implementation of GPP can be streamlined by using simple and standardized green criteria. Specifically, relying on internationally or nationally recognized standards such as ecolabels or the Greenhouse Gas Protocol supports the selection of suppliers and the assessment of environmental impacts associated with procurement decisions.[40] These standards can be adapted based on detailed market research to inform the availability and cost of green alternatives and can embed related requirements on inclusivity and equity (footnote 6).

A transparent monitoring and evaluation framework for procurement decisions and their associated environmental impacts ensures accountability and enables continuous improvement. Malaysia's GPP presents a good practice example. The Ministry of Finance is upgrading its e-procurement system to streamline GPP data tracking and results on GPP performance and total emission reduction achieved are regularly reported and monitored (Box A12).

Engaging stakeholders and implementing complementary policies are vital for fostering support and addressing potential supply constraints. Engagement with procuring agencies, line ministries, suppliers, and civil society is crucial for gathering feedback and securing buy-in for green procurement reforms (footnote 40). Complementary policies, such as publishing supplier road maps to net zero and supporting the development of green alternatives, can promote the availability of green products and services and facilitate implementation.

Ensuring a Sustainable Fiscal Trajectory

Context and Key Levers

CFAs manage fiscal trajectories to ensure alignment with climate objectives and sustainability in the face of climate risks. This requires reforming fiscal incentives to support economy-wide decarbonization and resilience objectives, while also rebalancing spending and revenue to maintain fiscal sustainability as the economy transitions and exposure to physical risks increases. A critical precondition for designing and implementing sustainable fiscal pathways is to embed risk assessments into budget management processes. Drawing on climate-related fiscal risk assessments, CFAs can consider how reforming fiscal incentives and rebalancing spending and revenue shapes these scenarios, including effects on physical and transition risks, revenues, and expenditure.

---

[39]   A. Blazey and M. Lelong. 2022. Green Budgeting: A Way Forward. *OECD Journal on Budgeting.* 22 (2). pp. 1–19; Output-based or results-based budgets are intended to hold budget managers accountable for their role in organizing the supply of goods and services to the public, and to enforce a regular review of the effectiveness of government expenditure programs.

[40]   H. La Cascia, et al. 2021. Green Public Procurement: An Overview of Green Reforms in Country Procurement Systems. *Climate Governance Papers.* World Bank.

The first set of actions involves reforms to fiscal policies to align with decarbonization and adaptation plans. The necessary reforms vary depending on the prevailing fiscal models and can include incentives for both mitigation (see below) and adaptation through measures such as water pricing or land use incentives. Three types of levers are particularly prominent across the region:

- **Phasing out of fossil fuel subsidies.** Widespread subsidies for fossil fuel, which in 2021 were worth 1%–3% of GDP in Indonesia, Malaysia, and Viet Nam have historically been used to support economic development and ensure energy affordability.[41] However, in practice, they incentivize emissions, thereby raising the cost of decarbonization, and are typically regressive—a study of 16 developing countries found 45% of the direct benefits of subsidies accrue to the top 10% of households by income, while another study estimates that reforms to explicit fossil fuel subsidies in 25 high-pollution, high-subsidy countries could save about 360,000 air-pollution-related deaths from 2002 to 2035.[42] The retirement of subsidies can therefore support mitigation objectives and create fiscal space—some of which may then be used for more targeted social support.

- **Carbon pricing.** This is a key fiscal lever to encourage decarbonization at a low cost but remains relatively underdeveloped across the region—among ASEAN member states, it has only been enacted directly in Indonesia and Singapore, and it is under consideration in different forms in Brunei Darussalam, Malaysia, Thailand, and Viet Nam.[43] Direct carbon pricing may take the form of tax or an emissions trading system, each with advantages and disadvantages, but both can create fiscal space. [44] Indirect approaches, where specific emission-intensive activities are relatively highly taxed, can have similar effects, though are typically less efficient than direct mechanisms in reducing emissions at low cost. Additionally, CFAs can explore the development of revenue streams through international carbon markets under Article 6 of the Paris Agreement or the voluntary carbon market (VCM), which can create export earnings for sectors such as agriculture to sequester carbon, as international buyers seek to offset emissions elsewhere. While in principle, VCMs could provide significant revenues and support emissions reductions, their efficacy is limited by the credibility of VCMs and there is an increasing convergence between Article 6 and VCM as guided by the Article 6 Rules.

- **Targeted incentives for low-carbon or resilient technologies.** Such incentives, which include investment grants, subsidized loans, and corporate income tax breaks, can complement broader-based carbon pricing by supporting investment in strategically important technologies. ASEAN member states have adopted these mechanisms for a variety of purposes, including the uptake of electric vehicles (EVs) (e.g., Indonesia's value-added tax reduction for EVs) or more investments in renewable energy (e.g., Malaysia's Green Investment Tax Allowance for green technology projects, Viet Nam's preferential tax rate for renewable projects). However, they typically imply increased government spending as well as revenue leakage (through revenue foregone as a result of tax incentives). As with fossil fuel subsidies, these can also be untargeted or regressive.

A second set of levers ensures that the fiscal trajectory is sustainable over the long term in the context of the transition to net zero and a changing climate. In the ASEAN region, countries such as Indonesia, Malaysia, and Brunei Darussalam generate over 15% of their annual government income from sources linked to fossil fuels, including taxes, export tariffs, and other nontax revenues, which would be expected to diminish as the use of these fuels declines.[45] Similarly, long-term impacts of physical risk on the value of agricultural production, which is expected to be reduced by approximately 10% in Indonesia alone by mid-century, could lead to significant shifts in spending (according to the Organisation for Economic Co-operation and Development [OECD], over 1% of GDP

[41]   IEA. 2023. *Fossil Fuels Consumption Subsidies 2022.*

[42]   F. J. A. del Granado, D. Coady, and R. Gillingham. 2012. The Unequal Benefits of Fuel Subsidies: A Review of Evidence for Developing Countries. *World Development.* 40 (11). pp. 2234–2248; and R. Damania, et al. 2023. *Detox Development: Repurposing Environmentally Harmful Subsidies.* World Bank.

[43]   Singapore introduced a carbon tax in 2019, and Indonesia implemented an emissions trading system for its power sector in 2023. The People's Republic of China, Kazakhstan, the Republic of Korea, and New Zealand have an emissions trading scheme (ETS), and Japan has a carbon tax and two sub-national ETS.

[44]   IMF. 2022. *Carbon Taxes or Emissions Trading Systems?: Instrument Choice and Design. Staff* Climate Note. No. 2022/066.

[45]   EU-ASEAN Business Council. 2023. Energy Transition in ASEAN. Singapore: EU-ASEAN Business Council; and D. Braithwaite and I. Gerasimchuk. 2019. Beyond Fossil Fuels: *Indonesia's Fiscal Transition.* Global Subsidies Initiative Report. International Institute for Sustainable Development.

was spent on agricultural support in 2022 in Indonesia alone).[46] Shifts on this scale may call for broad-based reform of tax and spending programs (footnote 6). Such reforms can include instruments that target rents or externalities associated with adaptation or mitigation programs: potential examples are land value tax to support investment in resilient or low-carbon infrastructure, export tariffs applied to green energy or carbon credits, road pricing for EVs and internal combustion engine vehicles, and royalties for minerals critical to the transition (with Indonesia having raised nickel royalties by 10% in 2019).[47]

Key Challenges

Ensuring reforms do not undermine social or development objectives or are politically infeasible. Significant shifts in taxation or spending policies are liable to create winners and losers. Particular risks associated with the phase-out of fossil fuel subsidies or carbon pricing include energy affordability, especially in countries that already face energy poverty, such as Cambodia, Indonesia, Myanmar, the Philippines, and Thailand, and impacts on the global competitiveness of emission-intensive exports.[48] Modeling evidence suggests that, while in the longer term, these impacts will be offset by behavioral change and productivity improvements, shorter-term impacts can be significant and negative.[49] Reforms can meet political resistance—for example, in 2022, a reduction of fossil fuel subsidies led to large protests in Jakarta and other cities—especially in settings where there is a lack of clarity on the intention of the reforms of the government's commitment to mitigating actions.[50] The lack of clarity of a government's commitment to reforms sometimes includes both the scope as well as the process. Resistance from various interest groups, for example, fossil fuel players, manufacturing industries that benefit from fossil fuel subsidies, or general discontent from the electorate may also discourage political leaders from going through with reforms and, in some cases, cause them to roll them back.

Avoiding incentives that are ineffective, inadequately targeted, or fiscally unsustainable. The deployment of specific incentives to support the uptake of decarbonization or resilience measures generally necessitates increased government spending—which may be unaffordable if uptake exceeds expectations. Other pitfalls associated with such measures include impacts that favor the wealthy—for instance, reduced excise taxes on EVs often disproportionately benefit higher-income households—or incentives that reward firms or households for actions they would have carried out without the incentive—for instance, in subsidizing rooftop solar panels where these are already cost-effective.

Ensuring that enabling infrastructure and regulatory frameworks are in place. Fiscal incentives alone may be ineffective, if not supported by adequate physical and regulatory infrastructure. For example, in the ASEAN region, there has been a focus on enhancing renewable power generation capacities, but these are likely to be ineffective unless the ASEAN invests in domestic and regional grids (footnote 4). Similarly, in Viet Nam, the effectiveness of renewable energy incentives has been compromised by inadequate regulatory frameworks, such as low selling prices under new schemes compared to previous feed-in tariffs.[51] While local banks have funded the considerable scale-up of renewable energy in Viet Nam in recent years, international lenders are needed to support further growth. Yet most international lenders have not been able to fund the sector on standard (i.e., nonrecourse) project finance terms due to significant shortcomings in risk allocations in the current power purchase agreement framework between the independent power producers and Vietnam Electricity, the offtaker. Regulatory frameworks may also be required to support the development of international carbon markets or allow investors in resilience infrastructure to monetize benefits that they provide (e.g., through insurance markets).

---

[46] OECD. 2023. *Agricultural Policy Monitoring and Evaluation 2023: Adapting Agriculture to Climate Change;* and ADB. 2021. *Climate Risk Profile: Indonesia.*

[47] M. Huxham, A. Muhammed, and D. Nelson. 2019. *Understanding the Impact of a Low Carbon Transition on South Africa.* A CPI Energy Finance Report, Climate Policy Initiative; and Reuters. 2019. Indonesia Nickel Ore Royalties Double in Regulations Shake-up. 10 December.

[48] K. Zhou, Y. Wang, and J. Hussain. 2022. Energy Poverty Assessment in the Belt and Road Initiative Countries: Based on Entropy Weight-Topsis Approach. *Energy Efficiency.* 15 (46). pp. 1–27.

[49] F. Venmans, E. Jane, and D. Nachtigall. 2020. Carbon Pricing and Competitiveness: Are They at Odds? *Climate Policy.* 20 (9). pp. 1070–1091; and ADB. 2016. *Fossil Fuel Subsidies in Asia: Trends, Impacts, and Reforms.*

[50] UNDP. 2021. *Fossil Fuel Subsidy Reforms: Lessons and Opportunities.*

[51] McKinsey & Company. 2023. Putting Renewable Energy Within Reach: Vietnam's High-Stakes Pivot. 2 October.

Designing reforms in the context of uncertainty over the nature and speed of climate-related shifts. The speed at which different sectors decarbonize and the impacts of physical climate risk are subject to significant uncertainty, meaning fiscal reforms that seek to balance revenues and spending over the medium term are likely to require revision. One particular risk is that premature migration of the tax base toward lower carbon activities (e.g., EV registration fees or royalties for critical minerals) could delay the overall transition to a low-carbon economy.

Good Practices

Reform Tax and Subsidy Systems to Incentivize Climate Mitigation and Adaptation Programs

To ensure reforms do not undermine social or development objectives, CFAs can prioritize reforms that do not carry significant negative impacts on sectors or groups of society. Win-wins of this kind are likely to be rare but may nonetheless be significant. As an example, the EU's CBAM comes into force in 2026 and will impose carbon pricing on products manufactured in ASEAN that are exported to the EU (footnote 7). The revenues from this can be onshored by CFAs imposing an export tax equal to the EU carbon tax on affected products since the CBAM does not impose double taxation.

For other policies, robust incidence assessments can support the design of fiscal reforms that mitigate negative impacts. Incidence assessments estimate impacts on incomes and employment across different sectors and groups of households, accounting for factors that determine vulnerability to shocks such as the ability to switch to other fuel sources. A good practice example is the incidence assessment conducted by the Indonesian Fiscal Policy Agency, in collaboration with PROSPERA, to assess the potential socioeconomic and fiscal impact of different options to reform the liquefied petroleum gas (LPG) subsidy (Box A13).[52]

Incidence assessments can support the development of reforms that mitigate negative impacts, for example through the following:

- **Phased approaches to avoid shocks to prices and incomes.** Phased approaches, which can include a gradual expansion of sectors covered by carbon prices or the ramping up or down of carbon prices or fuel subsidies, allow households and businesses to anticipate and adjust to the change in prices over time. The Government of Indonesia first targeted gasoline over kerosene when phasing out fossil fuels to avoid a more pronounced impact on low-income groups (footnote 52) (Box 14), while Singapore introduced its carbon tax with a clearly defined pricing schedule, starting with $5 per ton of carbon dioxide equivalent from 2019 to 2023 and planned price increases to reach $50–$80 per ton of carbon dioxide equivalent by 2030 (Box 15).

- **Targeted transition support mechanisms to mitigate adverse impacts and protect vulnerable groups.** These can include measures such as direct cash transfers or grants for the shift to cleaner technologies, or support to retrain workers so they can migrate to low-emissions sectors. In Indonesia, the 2005 subsidy reform made a provision (bantuan langsung tunai, or temporary unconditional cash transfer) to provide approximately $30 to low-income households (footnote 52). The Singaporean government adopted a mechanism to support Housing and Development Board households' purchase of energy and water. Malaysia recently implemented a differentiated approach to remove subsidies on diesel: industrial users are banned from pumping at the station, and low-income private diesel vehicle owners pay the full price at the pump but can apply for a monthly payout to help defray the cost.[53]

---

[52]   J. Kuehl, et al. 2021. *LPG Subsidy Reform in Indonesia: Lessons Learned from International Experience.* Global Subsidies Initiative Report. International Institute for Sustainable Development (IISD).

[53]   *The Business Times.* 2024. Malaysia Spending Shows Urgency of Fuel Subsidy Cuts. 2 July.

Consultative approaches to reforms linked with communication campaigns can make them more widely acceptable. Consultations can provide insight into how reforms impact groups, thus shaping incidence assessments and the detailed design of reforms to mitigate negative effects. Public campaigns can allow CFAs to communicate the rationale and benefits of reforms, an approach followed by Indonesia in fossil fuel subsidy reform, where positive impacts on long-term development were emphasized.[54]

Good practice in designing specific incentives ensures they are effectively targeted and minimizes their costs. This can include the following:

- **Clarity on the strategic rationale to inform the design and scope of the incentive instrument.** For example, if an incentive is intended to bring down the cost of a nascent technology or develop an infant domestic industry, its duration is likely to be time-limited, or subject to periodic market monitoring (footnote 6). If an incentive is intended to promote the uptake of technology by credit-constrained households, it can be subject to income or wealth thresholds.

- **The use of market mechanisms and efficient risk sharing to reduce costs.** For example, contracts-for-difference have been widely deployed to encourage renewable energy generation. These link subsidy payments to the difference between a pre-agreed upon "strike price" and prevailing wholesale prices, reducing risks to project developers, and can be subject to competitive bidding to reduce costs. Viet Nam has recently approved the use of a virtual direct purchase power agreement model with the national grid similar to a contract for difference. Under the arrangement, a corporate offtake agrees to a contract price and a contracted capacity with a generator, continues to buy electricity from Viet Nam Electricity at the market spot price, and settles the difference with the generator.

CFAs can take an active role in ensuring infrastructure and regulatory frameworks are in place. Following are good practices in this regard:

- **Shaping national transition road maps.** By driving the development of a clear transition or adaptation road map, the CFA can help the government formulate clear priorities, which can support the identification of any requirements for enabling conditions.

- **Close coordination of fiscal incentives.** The CFA can take on a central role in assessing the combined impact of different fiscal incentives and help line ministries coordinate their efforts. In particular, during the budget and investment planning process, CFAs can support line ministries in identifying required investments to unlock private climate adaptation and mitigation actions.

- **Collaboration to develop supporting policy framework.** CFAs can take an active and coordinating role in driving the development of supporting policy frameworks (e.g., feed-in tariffs for renewables). While CFAs are typically not tasked with designing new market policies, as the key authority in ensuring fiscal prudence, CFAs can challenge the introduction of incentives before exploiting other policy or regulatory changes that could improve the economics or feasibility of specific projects or programs.

Design Fiscal Reforms to Ensure Sustainable Public Revenue and Expenditure Trajectory

To address the technical challenge of assessing impacts across multiple possible scenarios, CFAs can simulate reforms using climate-related fiscal risk assessments. Building on the fiscal risk assessment, the CFA can, in more detail, explore expected impacts on government revenues, leveraging transition scenarios and economic development projections to explore the revenue generation potential of alternative fiscal levers. An example of this approach is the assessment of the fiscal implications of transition risk in South Africa, which scrutinizes how shifts in energy policy might affect national revenues (footnote 49). Other examples include interventions by CFAs to assess and engage in the design of fiscal incentives related to Malaysia's solar cell and module strategy (Box A16) and Thailand's EV road map (Box A17).

---

[54] J. Husar and F. Kitt. 2016. *Fossil Fuel Subsidy Reform in Mexico and Indonesia*. International Energy Agency.

To manage uncertainty, CFAs can establish a medium- and long-term plan for evolving the tax revenue system to accommodate new economic realities. CFAs could outline the future evolution of the tax system and define triggers for introducing specific tax reforms. This may involve investment in capacity (e.g., databases) in collaboration with the relevant line ministries and agencies. Regular monitoring of the long-term fiscal outlook can allow these trajectories to be periodically updated.

Building up Fiscal Resilience Against Climate-Related Shocks

Context and Key Levers

CFAs can reduce climate-related risks through the enactment of levers set out in previous sections but are still likely to face significant residual risks. In Southeast Asia, high and increasing climate risks can create significant fiscal instability. Estimates of the fiscal burden (expenditure and taxation) associated with a 1-in-200-year climate disaster can be as much as 20% of government expenditures, while transition risks can also cause systemic instability, for example, through the failure of critical industries leading to contagion through the financial system (footnote 2). While measures to allocate risks, support investment, and create efficient incentives can be important in reducing this volatility risk, it is neither feasible nor desirable to eliminate it over any foreseeable time horizon—and in countries with limited capacity to invest, it may not even be realistic to significantly reduce risks in the short term. As a consequence, CFAs in Southeast Asia are likely to face notable climate-related residual risks.

To manage residual risks, CFAs can deploy a mix of risk retention and transfer instruments:

(i) **Deploy risk retention instruments, including contingency funds and credit lines.** In Southeast Asia, countries have set up significant contingency funds to support disaster response: the Philippines, for example, has established both national and local disaster funds, the latter mandating local governments to allocate at least 5% of their annual budget to the Local Disaster Risk Reduction and Calamity Fund, while Indonesia and Viet Nam have instituted similar arrangements.[55] Contingent credit arrangements, such as ADB's Contingent Disaster Financing, can provide timely fiscal capacity at times of stress.[56]

(ii) **Deploy risk transfer mechanisms that can protect against the most extreme disasters.** Such measures can offer rapid payouts in the event of disasters in exchange for annual premiums and tend to be cost-efficient for low-probability, high-impact risks. Several Southeast Asian countries have adopted risk transfer measures for physical climate risks: in 2017, the Philippines secured parametric insurance that provided predetermined payouts for typhoons or earthquakes of specific magnitudes, underwritten by global reinsurers.[57] In 2021, the Lao PDR became a founding member of Southeast Asia Disaster Risk Insurance Facility (SEADRIF), Asia's inaugural risk financing facility, and procured parametric insurance against floods.[58] More recently, the Philippines participated in the World Bank's catastrophe (cat) bond program, under which the Government of the Philippines pays a fixed coupon to investors. In the event of predefined disasters, the bond's principal is not repaid to investors but directed to the government instead.[59] Other potential risk transfer instruments include loans with debt relief clauses activated by disasters.

---

[55] D. T. Villacin. 2017. A Review of Philippine Government Disaster Financing for Recovery and Reconstruction. *Discussion Paper Series.* No. 2017-21. Philippine Institute for Development Studies; World Bank. 2020. *Project Appraisal Document on a Proposed Loan in the Amount of US$ 500 Million to the Republic of Indonesia for the Indonesia Disaster Risk Finance and Insurance;* and T. T. L. Huong, et al. 2022. Disaster Risk Management System in Vietnam: Progress and Challenges. *Heliyon.* 8 (10).

[56] ADB. 2022. *Enhancing Contingent Disaster Financing and the Countercyclical Support Facility.* ADB Policy Paper.

[57] World Bank. 2020. *Lessons Learned: The Philippines Parametric Catastrophe Risk Insurance Program Pilot.*

[58] SEADRIF Insurance Company. 2024. Lao PDR Renews Parametric Flood Insurance from the SEADRIF Insurance Company. 2 February.

[59] OECD. 2024. *Fostering Catastrophe Bond Markets in Asia and the Pacific.* The Development Dimension Series; Indonesia has access to volcanic risk CAT bonds issued by the Danish Red Cross.

Key Challenges

Balancing risk retention instruments with risk transfer instruments. Risk transfer mechanisms, such as insurance and cat bonds, are typically expensive relative to the expected level of cover they provide, but, on the other hand, substantial risk retention requires building up liquid reserves at a high opportunity cost.[60] Balancing this trade-off, in the context of uncertain losses from climate, is a fundamental challenge.

Ensuring available funding matches the profile of needs. Immediate disaster response prioritizes the rapid availability of funds, while subsequent recovery and reconstruction phases prioritize support commensurate with the needs. This necessitates a well-coordinated approach that aligns the scope and timing of various instruments to ensure an effective and comprehensive fiscal response to disasters (footnote 60).

Approving and effectively deploying unfamiliar risk transfer mechanisms. For Southeast Asian countries exploring new forms of risk transfer such as parametric insurance, there are hurdles in gaining government and public support, which is critical for the successful adoption and implementation of these mechanisms. Developing a robust procurement process, supporting legal framework, and designing an effective disbursement process are all complex tasks that require expertise and careful planning. These challenges are compounded by the need to ensure that these mechanisms are both cost-effective and aligned with national disaster risk management strategies. A lack of familiarity and the intricate nature of these financial instruments can hinder their adoption and effectiveness iin managing financial risks arising from disasters triggered by natural hazards (footnote 60).

Good Practices

Deploy Risk Retention and Transfer Instruments

Good practices in managing these challenges apply jointly to decisions about risk retention and transfer instruments.

Balancing risk retention instruments with risk transfer instruments. The principle of risk layering can support a balanced portfolio of risk retention and transfer. This considers the frequency and severity of disasters and deploys risk retention measures for higher-frequency, lower-severity events (where the opportunity costs of budget reserves are lower), and risk transfer for higher-severity, lower-frequency measures (where transaction costs associated with insurance are lower), as illustrated in Figure 5. In the Philippines, this approach is implemented as follows (Box A7):

- For low-impact risk, the government established local and national disaster funds worth around $1.4 billion in 2021.[61]

- For moderate- to high-impact risks, the Philippines has secured $500 million contingent lines of credit from the World Bank.

- For severe, infrequent events, the government has piloted and set up specific risk transfer mechanisms, including parametric insurance that has disbursed $28 million over 2 years.[62]

---

[60] ADB. 2021. *Revised Disaster and Emergency Assistance Policy.*; and IMF. 2018. *How to Manage the Fiscal Costs of Natural Disasters.* How-To Note. No. 2018-003.

[61] Government of the Philippines, National Disaster Risk Reduction And Management Council (NDRRMC). 2022. *The Philippines' Midterm Review of the Implementation of the Sendai Framework for Disaster Risk Reduction 2015-2030 with a Short-term Review of the National Disaster Risk Reduction and Management Plan (2020-2030)*; ₱80 billion at an exchange rate of ₱57=$1.

[62] B. R. Signer and R. Poultier. 2021. Disaster Risk Insurance: 5 Insights from the Philippines *World Bank Blogs.* 1 December.

The development of an effective risk layering structure can be facilitated by the following:

- Using forward-looking methods to quantify the expected impact of disasters, integrating these in multiyear planning, and updating estimates based on changes in vulnerability (e.g., changes in insurance coverage). The Philippines, for instance, leveraged a country-specific catastrophe risk model with an improved asset exposure database for typhoon and earthquake risk for the development of their risk strategy and parametric insurance (footnote 60).

- Link risk management strategy to fiscal health indicators, ensuring that the risk management strategy supports the fiscal strategy. Accounting for both the current fiscal health and the potential fiscal health impacts of most extreme events can help identify priorities for strengthening fiscal buffers of transfer mechanisms.

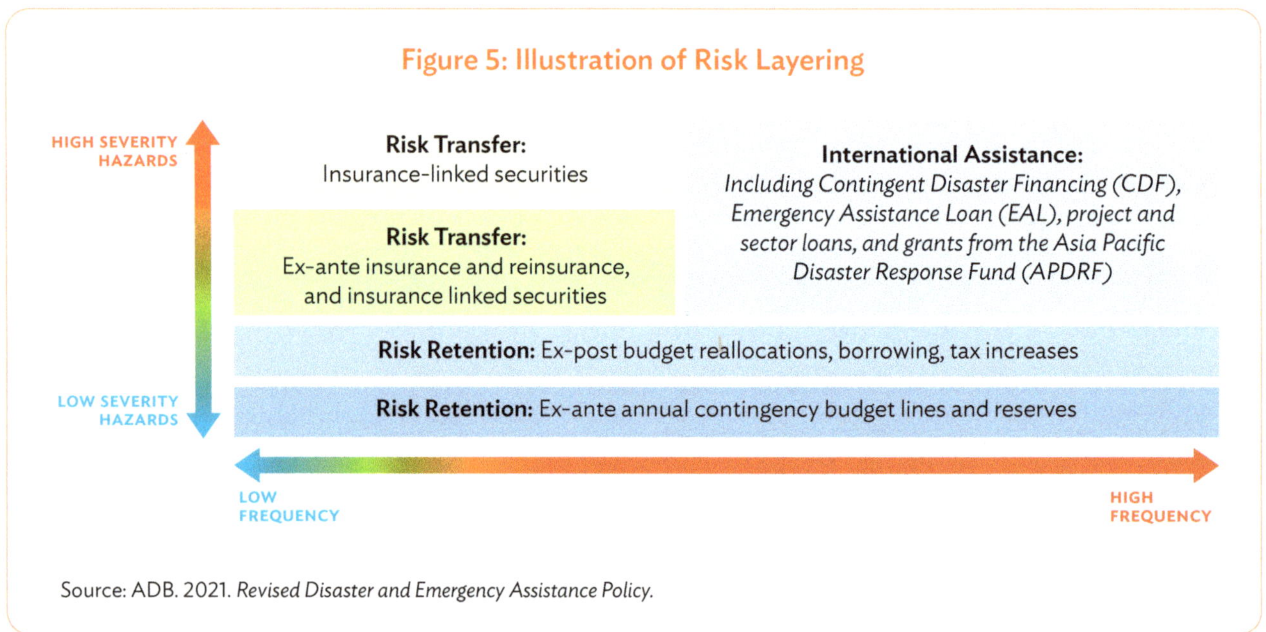

Figure 5: Illustration of Risk Layering

Source: ADB. 2021. *Revised Disaster and Emergency Assistance Policy.*

Ensuring available funding matches the profile of needs. Use a portfolio of instruments to provide phased support for disaster relief, recovery, and reconstruction. Mechanisms that allow rapid disbursement in the case of a disaster, such as parametric insurance, contingent credit lines, disaster funds, and national budget reserves, can be suitable for immediate relief funding, while other credit, insurance, or reconstruction funds can cover longer-term needs. The Lao PDR opted for an insurance policy that contains both a parametric- and indemnity-based insurance to ensure both rapid response and coverage of actual losses (Box A18).[63] Indonesia is likewise planning to combine different funding mechanisms (e.g., budget reserves, returns on invested reserves, parametric insurance) in its Pooling Fund for Disasters (PFB) to optimize its capacity to finance disaster relief and recovery (Box A19).

To ensure short-term disaster relief finance is effective, financing mechanisms can be paired with well-designed triggers and management rules. During the Philippines' parametric insurance pilot, disbursements were initially delayed due to a lack of clarity on the use of funds in the event of a payout (footnote 60). In the second year of the pilot, the payout rules were defined to improve the mechanism. This experience highlights the need for a clear definition of the payout triggers, the beneficiaries, and the use of funds, an approach that can assist in improving the planning of relief efforts as well as ensuring plans are funded.[64]

---

[63]   SEADRIF Insurance Company. SEADRIF Frequently Asked Questions.

[64]   D.J. Clarke and S. Dercon. 2016. *Dull Disasters? How planning ahead will make a difference.* Oxford University Press.

Approving and effectively deploying unfamiliar risk transfer mechanisms. A central technical body can provide capacity to set up and monitor disaster risk finance approaches. For instance, the Philippines established a central technical team to oversee the disbursement of the parametric insurance payouts. With the PFB, Indonesia is establishing a central entity that oversees national and subnational contingency funds and purchases appropriate risk transfer products. Two particular success factors for setting up disaster risk finance mechanisms in Indonesia, the Lao PDR, and the Philippines were the use of pilots and international collaboration:

- **Pilot approach.** The Philippines has piloted both parametric insurance and cat bonds with the technical support of the World Bank to explore the most suitable risk transfer mechanisms. Currently, the Philippines is even piloting an indemnity-based insurance for its public assets. Throughout the process, the Philippines further developed the technical expertise and capacity of a state-owned Government Service Insurance System, including the enhancement of its public asset database.

- **International collaboration and knowledge sharing: Indonesia, the Lao PDR, and the Philippines** developed the risk transfer mechanisms with the support of international partners (e.g., the World Bank and other national governments). The program development included significant knowledge-sharing formats to create strong ownership across different government agencies and to develop technical capacity.

4. Climate Finance Mobilization

This chapter reviews good practices to mobilize and deploy funding to support climate-related fiscal risk management and strategies. It covers public funds raised or mobilized from domestic revenue collection, sovereign debt issuance, and climate-related development support. It also reviews levers to enhance the capacity for public finance to support investment, expand access to public loans for climate-related investment, and align investments of sovereign wealth funds (SWFs) toward climate objectives. Finally, the chapter considers private funds mobilized from commercial financial institutions, institutional investors, and corporate investors, and discusses key levers to crowd in private capital and green the financial sector.

Mobilizing and Deploying Public Funds

Context and Key Levers

Significant resources are needed to fund climate-related investment, manage physical risks, and support people and businesses in transition. On green investment alone, the Monetary Authority of Singapore estimated that the Association of Southeast Asian Nations (ASEAN) member states will need $200 billion annually from 2016 to 2030.[65]

Public funds are critical to fund transition and adaptation projects. Despite the positive economic return on investment, adaptation and social infrastructure projects often lack cashflow potential, making them unattractive for private investors. Many climate-related infrastructure investments (e.g., resilient road networks) are inherently long term, requiring large upfront commitments and extended payback periods, while investments to deploy new technologies needed for the transition are risky, limiting appetite from private capital. Public funding may also be needed to provide affordable finance to vulnerable groups, such as farmers and small and medium-sized enterprises (SMEs), to enhance their climate resilience. Public funds are thus essential in transition and adaptation programs.

There are significant opportunities for CFAs in Southeast Asia to mobilize and deploy public funds for climate action. Revenue collection in Southeast Asia remains low (19.1% of GDP as of 2018) compared to OECD countries, and the Commitment to Reducing Inequality Index indicates that most Southeast Asian countries have the capacity to increase taxation progressively.[66] SOBs, such as Bank Rakyat Indonesia and Krung Thai Bank, play a major role in deploying loans that can support mitigation and adaptation projects. Additionally, several Southeast Asian countries have SWFs, including Singapore's GIC, Malaysia' Khazanah Nasional Berhad, Brunei Investment Agency, and Indonesia Investment Authority, which can channel resources toward climate objectives.

CFAs can explore three key levers to mobilize and deploy public funds:

(i) **Enhance capacity for public finance.** This can include: broadening the tax base through new forms of taxation (e.g., carbon and environmental tax) and tax reforms (e.g., reviewing existing tax treaties); improving tax collection and encouraging compliance; improving the efficiency of public spending and investment, and reducing revenue leakage, such as shifting away from regressive and poorly targeted subsidies or tax breaks; raising lower cost debt (within the sovereign's debt capacity, which may be very constrained) through sovereign bonds, green bonds, and other thematic bonds, and accessing climate-related concessional finance; reducing

---

65   Monetary Authority of Singapore (MAS). 2018. *The Asian Green Bonds Opportunity*. Opening Remarks by Mr Chia Der Jiun, Managing Director, MAS. IFC Green Bonds Asia: Opportunities for Financial Institutions Conference. 7 June.

66   Oxfam. 2020. *Towards Sustainable Tax Policies in ASEAN: A Case of Corporate Income Tax Incentives.*

borrowing costs through measures to improve transparency and accountability; and enhancing subnational finance to support local climate actions by cities and local authorities through measures including land value capture, property tax reform, and municipal green bonds (footnote 6).

(ii) **Green and expand access to public loans.** This can be done through SOBs, to improve access to finance for existing mitigation and adaptation projects as well as for underserved sectors and groups (e.g., affordable microfinance to farmers to improve climate resilience; project finance for large strategic coastal protection infrastructure).

(iii) **Green the investment of SWFs.** This can be done by mandating green investment, climate disclosure, and performance tracking.

Key Challenges

Negative consequences of tax increases. Tax increases can be unpopular and may deter foreign investment. For example, Singapore and Thailand have deferred the implementation of the OECD global minimum tax initiative, whereas Viet Nam had plans to partly compensate large foreign investors affected by the tax rate increase.[67]

Debt sustainability and debt servicing burden. According to ADB's public debt vulnerabilities ranking, which considers solvency, liquidity, fiscal stance, and financing needs, the Lao PDR was ranked highly vulnerable. It has been facing major challenges in maintaining debt sustainability amid mounting interest and pressure on its currency. In the same ranking, Cambodia, Indonesia, Malaysia, the Philippines, and Viet Nam were categorized as having moderate public debt vulnerabilities.[68] These current vulnerabilities limit fiscal space and constrain CFAs' capacity to raise further debt to finance climate-related spendings.

Lack of proactive and transparent debt management processes. The World Bank Debt Reporting Heat Map on Cambodia and the Lao PDR highlighted gaps in public debt reporting (e.g., information on last loans), debt management reporting (e.g., annual borrowing plan), and disclosure on contingent liabilities.[69] Difficulties in assessing climate-related risks, such as uncertainty surrounding liabilities arising from disasters, further complicate the process.

Constraints in the regulatory and legal framework, market infrastructure, and investor appetite pose further challenges in public borrowing. The absence of solid legal frameworks on subnational governments' borrowing rights and conditions and the associated disclosure requirement can hinder access to finance. Subnational governments' lack of credit history and experience could also limit lenders' appetite. Furthermore, the absence of enabling market infrastructure, such as credit rating agencies, can limit the development of public finance. For example, in Indonesia, subnational governments have not issued bonds and have mostly borrowed through on-lending from the Ministry of Finance, whereas Cambodia has not attempted to promote subnational government borrowing.[70]

Capacity constraints in SOBs and SWFs. Even when the external enabling conditions are in place, SOBs and SWFs with a green mandate may lack the capacity and technical capability to effectively integrate climate into lending and investment decisions. New climate technologies and solutions, such as offshore wind and nature-based adaptation projects, often present unfamiliar risks. Banks and SWFs may lack the expertise and tools to robustly assess the financial viability and potential risks associated with these investments.

---

[67] D. V. B. Khanh. 2024. Vietnam Goes Ahead With Global Minimum Tax But Considers Impact. *Bloomberg Tax*. 16 January; and *Reuters*. 2023. Vietnam Set to Raise Effective Tax Rate on Multinationals as Part of Global Deal. 27 November.

[68] B. Ferrarini, S. Dagli, and P. Mariano. 2023. Sovereign Debt Vulnerabilities in Asia and the Pacific. *ADB Economics Working Paper Series*. No. 680. ADB.

[69] Word Bank. 2022. *Debt Transparency: Debt Reporting Heat Map*.

[70] P. Smoke. 2019. Improving Subnational Government Development Finance in Emerging and Developing Economies: Towards a Strategic Approach. *ADBI Working Paper Series*. No. 921.

Lack of investment-ready projects. Even when the mandates of SOBs and SWFs explicitly support climate lending and investment, the dearth of viable projects could remain a bottleneck. This challenge is partially addressed by the function of CFAs in relation to climate investment (Chapter 3).

Good Practices

Enhance Capacity for Public Finance

Broadening tax base and improving collection. Measures to raise revenue can be combined with climate-related incentives. Examples of this include Viet Nam's introduction of an environmental protection tax on hazardous goods such as fossil fuels, coal, and plastic bags, and Singapore's introduction of a carbon tax in 2019, which is expected to generate S$1 billion in revenue in the first 5 years (Box A15).[71] Good practices related to such measures are described in Chapter 3.

International bodies have published guidance on broader measures to increase revenue collection. Fairness, equity, reciprocity, and accountability are key considerations to be considered in domestic resource mobilization, while reforms to tax collection can be built on enforcement, facilitation, and trust.[72] Such general reforms to taxation are not the focus of this report.

Raising sovereign debt and accessing development funding. CFAs can leverage climate-specific instruments to lower borrowing cost and debt burden. Sustainability-linked bonds (linked with specific key performance indicators) and green bonds (linked with specific projects or objectives) can offer a borrowing cost advantage and send a strong signal of government commitment to climate actions (footnote 6). Indonesia has successfully launched four issuances of green *sukuk* (Islamic bond) from 2018 to 2020 totaling $3.2 billion, which have gained very positive responses from investors (Box A20). Credit enhancement instruments from Climate Funds, MDBs, and IFIs, such as a partial credit guarantee, could be leveraged to improve access to and lower the cost of financing (footnote 6). Mechanisms such as debt–for–nature and debt–for–climate swaps could also be explored to reduce the debt burden.[73] Transparent and responsible practices in debt and public fund management are crucial to ensuring the effective use of borrowed capital and received funds. This involves measures such as ring-fencing proceeds, adhering to funding conditions of concessional finances, and identifying pipelines to ensure that the funds are allocated to catalytic projects.

Reducing and improving the efficiency of public spending. Implementing performance-based budgeting with a robust monitoring and evaluation framework can enhance spending efficiency. For example, the World Bank's recent public financial review suggested that Cambodia could get better value from its expenditure on public services through more strategic budget management, linking salary increases to performance, and better balancing of resource allocation across levels of government.[74] Such monitoring and evaluation practices could inform better resource mobilization, allocation, and disbursement, particularly for novel climate-related projects.

Focusing on reducing economically inefficient spending, such as shifting away from tax holidays or regressive subsidies, can free up significant fiscal space for some Southeast Asian countries. Tax incentives can impose significant fiscal costs without changing behavior.[75] The OECD Investment Policy Review of Southeast Asia estimates that such costs amount to 6% of GDP in Cambodia and 1% of GDP in the Philippines and Viet Nam.[76] By eliminating ineffective expenditure, countries can redirect resources toward climate actions.

---

[71] Dezan Shira and Associates. 2023. *An Overview of Vietnam's Environmental Protection Tax*. Vietnam Briefing. 14 November.
[72] R. Dom, et al. 2022. *Innovations in Tax Compliance: Building Trust, Navigating Politics, and Tailoring Reform*. World Bank.
[73] International Institute for Environment and Development (IIED). 2021. *Linking Sovereign Debt to Climate and Nature Outcomes. A Guide for Debt Managers and Environmental Decision Makers.*
[74] World Bank. 2024. *Cambodia - Public Finance Review : From Spending More to Spending Better.*
[75] S. James. 2014. Effectiveness of Tax and Non-Tax Incentives and Investments: Evidence and Policy Implications. Available at SSRN: https://ssrn.com/abstract=2401905.
[76] OECD. 2019. *OECD Investment Policy Review of Southeast Asia.*

Enhancing subnational finance. Establishing a clear framework for subnational debt management and monitoring is essential for ensuring the financial stability and responsible borrowing practices of local governments. This is particularly important given the significant role of local governments in meeting localized adaptation needs and responding to climate-related disasters. Such a framework can provide guidelines on debt ceilings, use of funds (e.g., use restricted to infrastructure investment), reporting requirements, etc. to foster fiscal discipline and facilitate better oversight by CFAs.

National-to-local lending is an effective way to build the credit history of subnational governments and to pilot credit-enhancing mechanisms. These programs provide local authorities with access to necessary funds for climate-related investment while allowing them to build credit experience that will facilitate future borrowing (footnote 6). For example, the Regional Infrastructure Development Fund in Indonesia lends directly to subnational governments to help address their critical infrastructure needs. A fiscal transfer intercept mechanism was also piloted as part of this fund, which was then institutionalized to enhance the credit profile of subnational government borrowings (Box A21). Technical support in pipeline development, transaction execution, and debt management could further enhance subnational governments' capacity to undertake borrowings effectively and responsibly.

Green and Expand Access to Public Loans

Setting clear "green" definitions and lending targets. Establishing a green taxonomy helps financial institutions identify and categorize projects, supporting the effective channeling of funds into initiatives that promote environmental goals. In the absence of a green taxonomy, the preferential credit program by Agribank in Viet Nam established a clear set of criteria for "high-tech and clean" agriculture projects to support lending decision-making (Box A22).

Ensuring complementarity across capability-building measures. These include budget support for management and staff training on project appraisal, risk assessment, governance, etc., which are also vital for successful implementation. Robust project assessment and risk management are crucial for ensuring that new lending activities, such as offshore wind project financing, do not expose SOBs to unfamiliar risks. Additionally, strong governance and disclosure practices are crucial for monitoring performance and tracking progress.

Leveraging innovative debt instruments to fund green lending activities. The pilots by SOBs can demonstrate the feasibility of financing options such as green bonds and social bonds, and lay the groundwork for their broader deployment. One example is the $295 million social bond issued by Thailand's Government Savings Bank (GSB) to provide low-interest loans to grassroots populations and support the upskilling of vulnerable groups (Box A23). Although GSB does not explicitly target climate-related social investment, a similar approach could be used to fund a just transition.

Green the Investment of Sovereign Wealth Funds

Provide clear investment guidelines, complementary capability building, robust disclosure, and aligned incentives. Simple guidelines, such as product-based and conduct-based exclusion criteria, along with capability-building efforts, such as budget support for management training, can facilitate the successful implementation of the new green mandate. For example, GIC in Singapore has set up dedicated teams to capture sustainability-related opportunities (Box A24). Aligning management incentives with the new climate-related targets further reinforces the commitment to sustainable investment, while a robust disclosure framework enhances transparency and accountability. These can be embedded alongside wider good practices (including effective oversight from fiscal authorities and anticorruption measures) to ensure that public resources are used efficiently.

Mobilizing and Deploying Private Capital and International Funds

Context and Key Levers

Given the sheer scale of investment required for transition and adaptation programs, private capital is critical to ensure sufficient resources are available to fund green investment. As well as providing financial capacity, the private sector can bring greater efficiency, innovation, and technical know-how to funded projects. However, the supply of bankable projects remains low, reflecting a lack of cashflow potential for many projects or adverse risk–return profiles. Marsh and McLennan estimate that at least 35% of infrastructure projects in emerging markets in Asia are marginally bankable, with a further 55% being classified as unbankable.[77]

Private climate finance is still at an early stage of development in Southeast Asia. Many member states continue to rely heavily on public funding to meet their NDC commitments, and private sector participation is uneven across the region.[78] Blended finance initiatives have primarily focused on renewables, with limited activity in areas such as early coal retirement, climate tech, and the decarbonization of hard-to-abate industrial sectors.[79] There is also limited private capital activity in adaptation and social infrastructure investment for just transition.

CFAs could explore two key levers to mobilize private capital:

(i)   **Crowd in private capital.**

    (a) Shape blended finance initiatives through partnerships with development finance institutions. The initiatives could take the form of a consolidated financing platform (e.g., Sustainable Development Goal [SDG] Indonesia One [SIO] platform), a green finance facility (e.g., ASEAN Catalytic Green Finance Facility [ACGF]), or de-risking instruments offered in specific transactions (e.g., guarantee, grants for project preparation, equity investment.

    (b) Forge an enabling regulatory environment and providing investment incentives to support and attract capital flow. Supportive regulations directly within the realm of CFAs could include measures such as improving the ease of profit repatriation and simplifying tax filing processes. Incentives to attract foreign direct investment could take the form of capital grants for equipment or time-bound import duty exemption.

(ii) **Systemically green the financial sector.** CFAs can shape financial regulations to lay the foundation for climate finance deployment. Potential actions include developing a green finance road map and establishing a green taxonomy to guide investments; developing regulations on climate risk assessment, disclosure, and stress testing to promote transparency; and offering incentives to both banks and borrowers to facilitate the rollout and adoption of climate finance initiatives.

Key Challenges

Shallow domestic capital market and limited foreign investor appetite. The Asian financial sector is characterized by short-term bank lending and the lack of long-term investors, such as insurers and pension funds.[80] The McKinsey Asian Capital Markets Development Index notes challenges in funding at scale, investment opportunities, and pricing efficiency in countries such as Viet Nam, Indonesia, and the Philippines.[81] Investors' appetite is further limited by country risk considerations such as political stability and the strength of the legal framework.

---

[77]   Marsh & McLennan Companies. 2017. *Closing the Finance Gap: Infrastructure Project Bankability in Asia.*

[78]   G. Krishnan, et al. 2024. Accelerating Public Sector Support for Climate Financing in ASEAN. *ASEAN Socio-Cultural Community Policy Brief* No. 3.

[79]   Monetary Authority of Singapore (MAS). 2024. *Tiny but Mighty – Scaling Blended Finance in Asia.* Presentation at the Philanthropy Asia Summit. 17 April.

[80]   A. Sheng, C. S. Ng and C. Edelmann. 2013. *Asia Finance 2020: Framing a New Asian Financial Architecture.* Marsh & McLennan Companies.

[81]   McKinsey & Company. 2017. *Deepening Capital Markets in Emerging Economies.*

Policy uncertainty. Despite ambitious decarbonization targets in some Southeast Asian countries, there is often limited clarity on sector transition pathways to achieve these targets. Embedded policies favoring high-emission alternatives (e.g., fossil fuel subsidies) and limited clarity on the position regarding coal assets further undermine investor confidence. Additionally, regulatory barriers, such as lengthy and complex permitting processes for offshore wind and the uncertainty around power purchase agreements further deter investors.[82]

Limited enabling market infrastructure and data availability. The lack of enabling financial market infrastructure, such as rating agencies and risk transfer mechanisms, presents challenges for potential investors. They may struggle to obtain accurate and reliable information to inform investment decisions and lack tools to hedge against climate-related uncertainties. Additionally, the lack of robust physical and transition risk data presents significant hurdles to project appraisal, portfolio monitoring, and reporting by potential lenders and investors.

Lack of experience and capacity in climate finance. Specialized expertise is required to structure blended finance transactions to address barriers faced by transition or adaptation projects. For example, a solar project could face curtailment risk due to limited grid capacity, a coal plant early retirement project could face financial risks given the substantial costs involved, and a nature-based flood management project could face performance risks due to the dynamic design.[83] The level of financial and technical expertise required to successfully design and execute effective financing structures could be a significant hurdle.

Good Practices

Crowd in Private Capital

Thorough diagnostics on private investment barriers can be the first step to inform enabling policy actions and financial instrument offerings. Given CFA's central vision on investment and spending programs, CFAs could lead the diagnostics work to identify key issues that hinder capital flow and inform targeted policy actions. This process could also be leveraged to initiate dialogues with other line ministries to create necessary enabling conditions. For instance, ADB highlights that NAPs in many developing countries in Asia rarely provide the necessary level of detail for investment programs, which deters investors from deploying finance at scale.[84] Diagnosing the hurdles underlying these investment barriers can be CFAs' first step to mobilizing private capital.

CFAs can deploy risk-tolerant capital in a targeted way to make projects bankable. . For example, the Sustainable Development Goal Indonesia One (SIO) platform leverages four different types of facilities, covering project preparation grants, guarantees, senior and subordinated loans, and equity investment to de-risk and enhance project bankability (Box A25). Central platforms that consolidate capital providers with different risk-return preferences facilitate efficient matching between capital and project-specific funding gaps. The Financing Asia's Transition Partnership (FAST-P) blended finance partnership among MAS, ADB, and the Global Energy Alliance for People and Planet (Box A26), along with the SIO platform, exemplifies good practices in combining and channeling public and private finance to achieve climate goals.

Close collaboration with DFIs and international partners can combine finance with requisite technical capacity. The Cambodia National Solar Park Project is an example of combining technical support, concessional loans, and grants from partners such as ADB, the World Bank, and The Republic of Korea e-Asia and Knowledge Partnership Fund to initiate solar power development (Box A27). There are also meaningful opportunities to leverage targeted facilities and mechanisms developed by DFIs to tackle transition and adaptation project challenges. For example, ADB's Energy Transition Mechanism (ETM) concept has been piloted by ACEN in the Philippines to accelerate coal plant retirement while protecting local communities and unlocking investment into green energy. The ASEAN Catalytic Green Finance Facility (ACGF) has also been utilized by member countries to finance infrastructure projects that promote environmental sustainability and contribute to climate change goals, such as the Agricultural Value Chain Competitiveness and Safety Enhancement Project in Cambodia (Box A28).

---

[82]   OECD. 2024. *Clean Energy Finance and Investment Roadmap of the Philippines.*

[83]   S. Janssen, et al. 2020. On the Nature Based Flood Defence Dilemma and Its Resolution: A Game Theory Based Analysis. *Science of the Total Environment.* 705.

[84]   ADB. 2021. *Climate Adaptation Investment Planning: A Program to Bridge the Gap between Climate Adaptation Planning and Financing.*

Systemically Green the Financial Sector

Regulatory reforms can adapt existing standards to local conditions, in consultation with the financial sector. Aligning disclosure requirements and taxonomy with regional and international standards ensures cross-border interoperability and facilitates access to foreign capital. For instance, Indonesia's green taxonomy adopts the foundation framework from the ASEAN Taxonomy and color-codes activities based on their impact on climate mitigation and adaptation projects (Box A29). Close consultation with financial institutions, investors, and international partners could allow CFAs to leverage shared experiences and avoid potential pitfalls. Clear communication of initiatives, such as the publication of a green finance road map for the financial sector, signals policy commitment and provides guidance for financial institutions and investors (footnote 6).

Complementary support for capacity building can equip financial institutions with the necessary tools and resources to take the envisioned climate finance initiatives off the ground. Climate finance initiatives often entail additional costs, including updates to banks' project appraisal frameworks, risk assessment methods, and internal monitoring systems. Singapore has addressed these frictions by introducing a grant scheme covering expenses related to the development of green and sustainability-linked loan frameworks (Box A30). Banks such as BNP Paribas, OCBC Bank, and UOB have since launched green loan frameworks that qualify for the scheme.

5. Conclusions

This report advances a central role for CFAs in climate policy covering both the management of fiscal exposures to physical and transition risks and broader support for national adaptation and decarbonization objectives. It highlights the relevance of a range of levers to countries in Southeast Asia, from understanding risks to allocating risks among public and private actors to managing fiscal incentives and developing investment strategies to unlocking required flows of public and private finance, highlighting good practices from countries across the region (Figure 6).

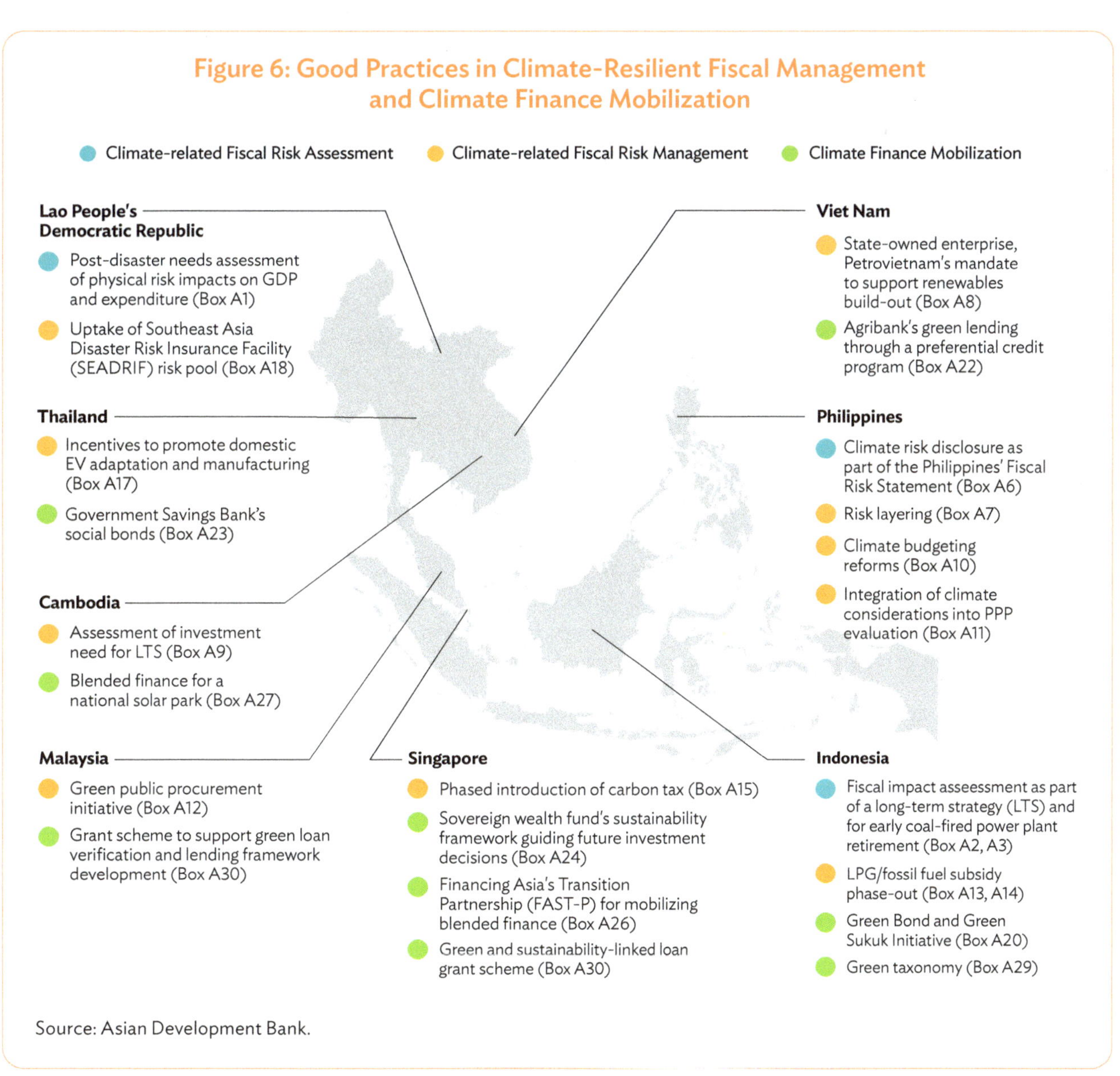

Southeast Asian countries have varying levels of resources, and will thus need to prioritize its actions for climate-resilient fiscal management and climate finance mobilization accordingly. This final section concludes with observations on key priorities for CFAs at different stages of implementing the practices reviewed in this report (Figure 7).

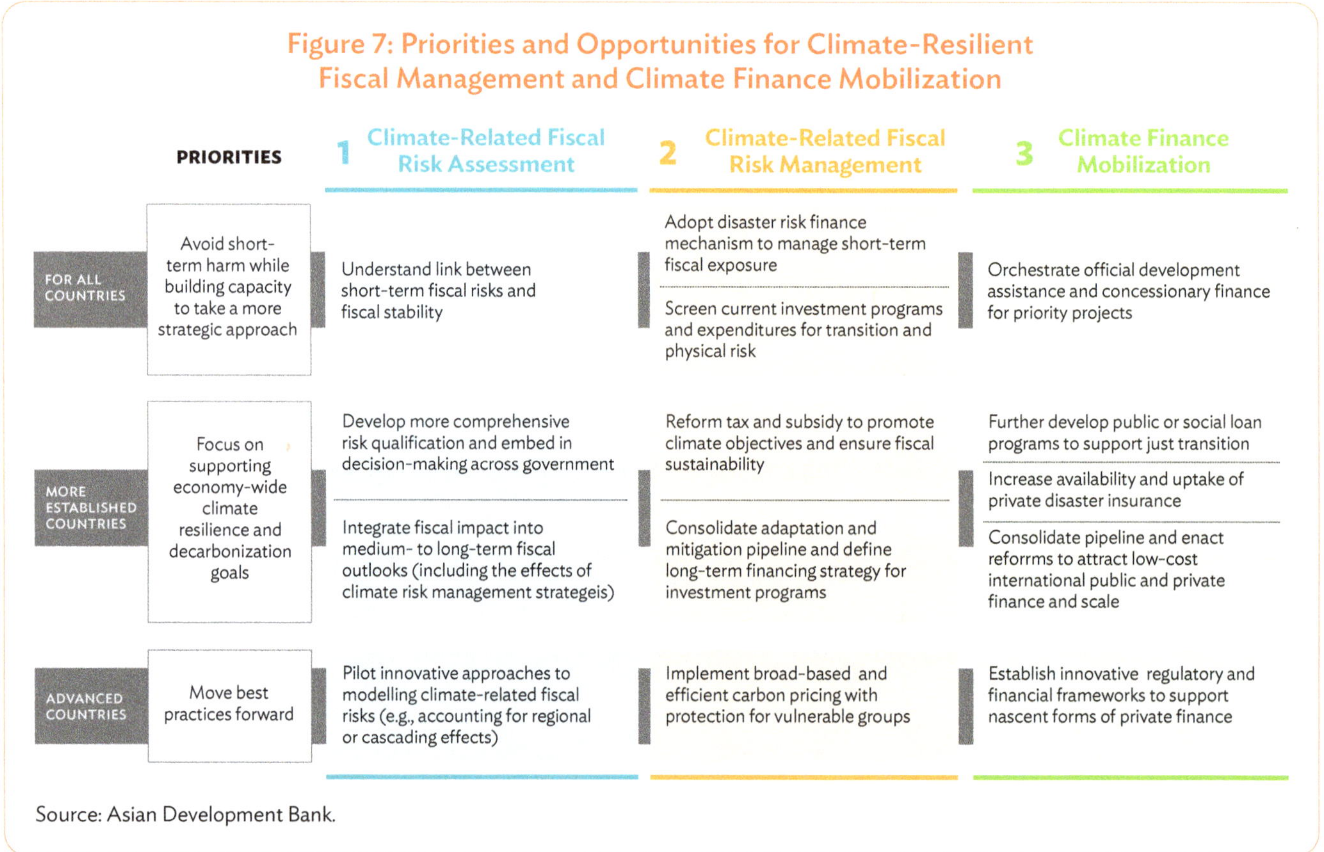

Figure 7: Priorities and Opportunities for Climate-Resilient Fiscal Management and Climate Finance Mobilization

Source: Asian Development Bank.

Short- and medium-term priorities for CFAs in Southeast Asian countries will depend in part on how advanced they are on the journey of adopting good practices. While all the steps of the framework can be pursued synergistically, CFAs may prioritize interventions based on the degree to which good practice is already established, as well as their specific exposures to physical and transition risk.

An initial step for all CFAs to take is to avoid short-term harm while building capacity to take a more strategic approach. This can involve actions across all three steps of the framework developed in this paper:

- **Climate-related fiscal risk assessment.** CFAs can begin by screening potential climate-related fiscal risks for short-term materiality and highlighting these to line ministries whose decisions have the greatest bearing on how these risks are managed. The assessment can draw on existing reports that prioritize decarbonization or adaptation interventions (for example, NDCs and NAPs) as well as international evidence on risks, structuring the review using established risk frameworks (footnote 6). Targeted quantitative modeling of impacts—potentially covering the fiscal consequences of extreme climate events on tax bases and implicit and explicit liabilities—can support the business case for related management decisions.

- **Climate-related fiscal risk management.** CFAs can focus on the most pressing risks identified in the first stage. For Southeast Asian countries that generally face high exposure to physical risks, a short-term priority is likely to be to reduce their destabilizing effects on public finances and the broader economy through the adoption of disaster risk finance strategies, supported by clarity of risk ownership between government agencies and the private sector. CFAs can look to disaster risk financing and management approaches adopted in the Philippines, which has contingent access to hundreds of millions of dollars to support relief and recovery, and the Lao PDR, which has taken up a parametric flood insurance policy through the SEADRIF Insurance Company. In addition, to avoid short-term decisions that lock in high physical or transition risk, CFAs can assess sectoral programs for climate risk, including implications for the fiscal outlook, and change course where required.

- **Climate finance mobilization.** CFAs can focus on ensuring public funds are deployed at minimum cost, working where appropriate with official development assistance partners to access concessionary finance by ensuring projects meet lender or donor requirements.

CFAs with more established approaches can focus on supporting economy-wide resilience and decarbonization goals.

- **Climate-related fiscal risk assessment.** CFAs that already have a high-level understanding of the macrofiscal importance of climate risk can work to develop this into a more systematic assessment of fiscal impacts. Following the example of Armenia, this could set out pathways for key metrics of fiscal health (e.g., debt-to-GDP ratios) out to the middle of the century, using scenario assumptions from widely recognized sources such as NGFS or IPCC. Such information can support decisions across government and form the basis of a fiscal risk disclosure to investors.

- **Climate-related fiscal management.** A more detailed assessment of risks can support the development of balanced fiscal pathways promoting adaptation and decarbonization objectives. This will involve detailed investment plans across all key sectors and potentially include the phased implementation of specific incentives such as carbon pricing or subsidy reforms, with measures to protect vulnerable groups. Such protections may take similar forms as in Indonesia's fuel subsidy reforms, which offered targeted disbursements worth $30 to low-income households.

- **Climate finance mobilization.** With detailed investment programs set out, a key priority for CFAs is to develop credible financing strategies, working with DFIs, investors, and the private sector to access finance at an efficient scale, making use of blended approaches and climate finance instruments where appropriate. The implications of the finance strategy for long-term fiscal sustainability and stability can be incorporated into the fiscal risk assessment, with assumptions on access to and cost of finance disclosed to investors.

Finally, the most advanced CFAs can advance the frontier of best practice. Areas of focus for this may include more advanced modeling of the relationship between climate and fiscal risks, for instance, accounting for cascading effects of risks along value chains and through the financial system. More ambitious policy models applicable to Southeast Asian countries could include revenue raising through international carbon markets or incentive regimes for sustainable agriculture, while new financial models could support more blended finance for adaptation or create fiscal capacity through debt–for–nature swaps. By developing advanced tools that can be applied elsewhere, these CFAs can play a catalytic role across the region. Singapore, for example, has established the FAST-P to mobilize $5 billion that will support not only local but regional needs for investment.

To move forward, CFAs can use the framework to define priorities and build programs of action. An initial stocktaking exercise can highlight gaps between current policy frameworks and good practices set out in this report. This, in turn, can inform strategic priorities, allowing CFAs to define programs of reforms and identify resources required for their enactment. For many CFAs, these resources will include partnerships with external bodies such as ADB that can provide finance, technical assistance, and capacity building.

Appendix I. Case Studies: Climate-Related Fiscal Risk Assessment

Climate-Related Risk Impact Assessment

Box A1: The Lao PDR's Post-Disaster Needs Assessment Following the 2018 Floods

In 2018, the Lao People's Democratic Republic (the Lao PDR) experienced record-high rainfall, leading to severe flash flooding that affected around 127,000 households nationwide. In response, the government conducted a post-disaster needs analysis (PDNA) immediately after the flood. It aimed to streamline disaster recovery and improve post-disaster management and reconstruction plans by assessing the impacts on major economic sectors and populations. To build the foundation for a comprehensive disaster recovery framework, the government asked the international community to assist with the PDNA on 3 September 2018.[a] The PDNA conducted a detailed assessment of the direct damages, economic losses, and need for recovery by comparing economic conditions before and after the disaster. The economic losses were estimated based on the changes in economic flows caused by the absence of damaged assets or disrupted access to goods and services, leading to revenue losses, higher costs, etc.[b] Moreover, the assessment estimated the impact of damages and associated economic losses on key fiscal indicators, such as

- increase of budget deficit to 4.7% instead of 4.3% of gross domestic product (GDP), and an additional increase of 0.9% in the following years;

- redirection of 1.5% of expenditures for medium-term recovery;

- sizing of impact on public assets in terms of damages and losses (mainly driven by infrastructure assets); and

- impact on future real GDP growth—the decline of projected growth projections by 0.2 percentage points.

The prerequisites for the effective implementation of this approach are (i) availability of data to estimate direct damages and economic losses; and (ii) substantial collaboration between government agencies, private sector partners, civil society organizations, and development partners.

The assessment enabled the Lao PDR to understand the impact channels and magnitudes in a timely manner and define a targeted strategy for pivotal recovery areas.

Good practice element and discussion on applicability:

- **Include second-order economic impacts and impacts on fiscal health indicators in the scope and coverage of climate risk impact assessment.**

---

[a] Government of the Lao PDR. 2018. Post-Disaster Needs Assessment 2018 Floods, Lao PDR.
[b] Global Facility for Disaster Reduction and Recovery (GFDRR). 2013. *Post-Disaster Needs Assessment Guidelines and Recovery Planning.*
Source: Asian Development Bank.

Box A2: Fiscal and Economic Impact Assessment as Part of Indonesia's Long-Term Strategy

In 2021, the Government of Indonesia published the Long-Term Strategy (LTS) for Low Carbon and Climate Resilience 2050. To formulate the strategy, the Government of Indonesia analyzed the economic impact of specific mitigation actions.

The government used modeling capabilities to assess the economic and fiscal impacts of its LTS:

- The economic impacts were assessed using the Asia-Pacific Integrated Model/Computable General Equilibrium-Indonesia, a multisector, recursive dynamic model that can project the economic and environmental impact of any policy implementation at the national level. The model enabled the assessment of the impact of climate mitigation actions in the agriculture, forestry, and other land use, waste, industry and energy sectors. Beyond economic growth, the assessment estimated an expected increase in government expenditures, national investment levels (not differentiating between private and public investments), and an impact on the employment rate (which could be linked to expected tax revenue).[a]

- The assessment deployed different development scenarios as well as three mitigation scenarios defining various levels of change in land use and energy generation.

The prerequisites for the effective implementation of this approach are (i) a politically defined and supported transition pathway to inform scenario selection and (ii) macroeconomic modeling capabilities.

Good practice elements and discussion on applicability:

- **Include fiscal health indicators in the scope and coverage of climate risk impact assessment.**
- **Conduct scenario modeling or stress testing.**

---

[a] Government of Indonesia. 2021. Indonesia Long-Term Strategy for Low Carbon and Climate Resilience 2050. United Nations Framework Convention on Climate Change.

Source: Asian Development Bank.

Box A3: Fiscal Impact Assessment for Early Coal Plant Retirement in Indonesia

Indonesia plans to retire coal-fired power plants (CFPPs) from 2031 to 2035. In Indonesia, 62% of electricity is generated by CFPPs and 35% of coal plants are less than 8 years old. In 2020, the coal industry contributed to approximately 4% of national fiscal revenue, while for coal-dependent provinces, this share amounted to 7%–34%.[a] During COP26, the minister of Energy and Mineral Resources stated that Indonesia will start CFPP retirement in 2031–2035. Therefore, it is critical for the Government of Indonesia to understand and compare fiscal impacts associated with the early retirement of coal assets (e.g., lower electricity subsidies, lower carbon tax collection, higher revenues from coal exports, etc.).[b]

The government assessed the potential impacts of the early retirement of existing coal assets.[c] Together with the energy transition partnership, the Ministries of Finance, Energy and Mineral Resources, and National Development Planning have started assessing the financial implications of early CFPP retirement, paying careful attention to the impact on the national fiscal position, including

- revenue impact from reduced tax (e.g., corporate income tax, income tax, and carbon tax) and increased tax and royalties from other sectors (e.g., higher export revenues for coal and renewables installation); and

- spending impact from increased compensation and social benefits to support a just transition and workforce reskilling and upskilling, as well as net change in spending due to reduced domestic coal used and increased electricity subsidy for renewable energy generation.

The prerequisites for the effective implementation of this approach are (i) an in-depth assessment of relevant tax and subsidy systems (e.g., Indonesia has 12 relevant support mechanisms for the electricity sector), (ii) a transition pathway developed to inform scenario selection (and scenarios also informing the transition pathway), and (iii) the need to support the assessment of the impacts on energy security and affordability.

Two scenarios were analyzed: a baseline scenario with no early retirement and a net-zero emission scenario in which CFPPs are retired and replaced by renewable energy by 2050.

Good practice elements and discussion on applicability:

- **Include fiscal health indicators in the scope and coverage of climate risk impact assessment.**
- **Conduct scenario modeling or stress testing.**

---

[a] Climate Bonds Initiative. 2022. Asean Economies Exposure to Climate Transition Risks.
[b] ASEAN Taxonomy Board. 2023. *ASEAN Taxonomy for Sustainable Finance*.
[c] The United Nations Office for Project Services and Energy Transition Partnership. 2022. Study on the financial implications of the early retirement of coal-fired power plants in Indonesia.

Source: Asian Development Bank.

Box A4: Assessment of Indonesia's Fiscal Risk Resulting from Carbon Pricing in the People's Republic of China

The People's Republic of China (the PRC) is the main importer of Indonesian coal and has an ambitious low-carbon transition agenda. The PRC has introduced a national emissions trading system (ETS) alongside regional pilot ETS in its regions and supports the development of renewable energy sources through subsidies and regulations. To achieve these decarbonization targets, the PRC may need to decrease its coal imports from Indonesia.

A technical paper by the Task Force on Climate, Development and the International Monetary Fund, supporting the Intergovernmental Group of Twenty-Four, analyzed the macrofiscal relevance of transition spillover risk in Indonesia, providing a quantitative assessment of its impact on fiscal and financial stability.[a] It focused on the implications of the Network for Greening the Financial System (NGFS) scenarios of change in Chinese demand for coal on Indonesia's macroeconomic performance (e.g., gross domestic product [GDP], unemployment, balance of payment) and sovereign financial stability (e.g., debt–to–GDP ratio). Studying the impact of spillover risk on the adjustments in the balance of payments and debt–to–GDP ratios is important for informing financial supervisory work. The analysis tailors and applies the EIRIN macroeconomic model to identify and quantitatively assess the shock transmission channels to agents and sectors of the economy and finance in Indonesia, as well as the drivers of shock amplification and spillover effects.

The assessment draws on three scenarios produced by the NGFS in 2021:

- **Current policies.** This scenario assumes that only currently implemented policies are preserved and emissions will grow until 2080, leading to about 3°C warming ("business-as-usual" scenario).

- **Below 2°C.** This scenario assumes that the stringency of climate policies gradually increases, starting immediately, creating a 67% chance of limiting global warming to below 2°C.

- **Net-zero 2050.** This scenario limits global warming to 1.5°C by the end of the century through stringent climate policies and innovation.

The assessment compares the impact of the NGFS's climate transition scenarios with the impact of spillover risk on key macroeconomic and public finance indicators. It focuses on the spillover risk's direct (shrinking export markets) and indirect (i.e., asset prices, investment, and fiscal revenue) impacts in Indonesia.

The prerequisite for the effective implementation of this approach is modeling capabilities on sector transition pathways, international trade, and macrofiscal consequences.

The results show that Indonesia is exposed to climate transition risk via spillover risk, both directly and indirectly. The spillover risk can induce trade-offs in sovereign economic and financial stability and decarbonization. On the one hand, spillover risk negatively affects GDP growth and the main macroeconomic indicators in Indonesia. On the other hand, it leads to lower greenhouse gas emissions in Indonesia, but the decrease is not enough to achieve the country's ambitious climate mitigation targets..

Good practice elements and discussion on applicability:

- **Include fiscal health indicators in the scope and coverage of climate risk impact assessment.**
- **Conduct scenario modeling or stress testing.**

---

[a] R. Gourdel, I. Monasterolo, and K.P. Gall. 2022. *Climate Transition Spillovers and Sovereign Risk: Evidence from Indonesia.*
Source: Asian Development Bank.

Box A5: Armenia's Comprehensive Risk Identification and Fiscal Impact Assessment

Armenia is highly exposed to acute and chronic climate-related impacts threatening the economy and society. Relevant acute risks include droughts, landslides, mudflows, and wildfires, while chronic risks include increased average temperatures, decreased precipitation, and land degradation.[a]

In 2022, the Ministry of Finance started including disaster and climate risk into its fiscal risk statement. After identifying acute and chronic risks and providing a qualitative analysis of their impact on the macrofiscal environment, the ministry worked with the International Monetary Fund (IMF) to quantify the risks, using an econometric model to reflect simulations in an analytical framework that links estimates of changing climate patterns on the real economy and with long-term fiscal projections. This considered: (i) the impact of higher temperatures on the economy, based on an empirical analysis of the effect of past temperature changes on growth; (ii) how slower economic growth flows through to fiscal projections to identify increasing fiscal pressures; and (iii) discrete climate change-related fiscal risks—including those translated through state-owned enterprises (SOEs) and public–private partnerships (PPPs)—to identify the state's direct exposures to climate risks. The study assessed the fiscal pressure by focusing on fiscal health and economic growth indicators, such as gross domestic product (GDP), public expenditures, net borrowing, and the debt–to–GDP ratio.[b]

The study developed the following three scenarios which projected the long-term fiscal situation relative to a baseline where current policies are followed and there are no additional physical risks.

- The Paris Agreement scenario, where the country (and world) meets its international commitments under the agreement. In this scenario, GDP remains unchanged from the baseline and public debt is 46% of GDP by 2072.

- The unmitigated scenario, where global greenhouse gas emissions continue to increase throughout the century, leading to a temperature increase of about 5.5°C above average 1990s levels by 2090–2100. Under this scenario, public debt levels could reach 62% of GDP if there is no fiscal policy response. Investing in adaptation can reduce the fiscal impact, even in the unmitigated scenario, and reduce public debt from 62% to 54% of GDP by 2072.

- The volatile scenario takes the unmitigated scenario and adds the volatility of weather and more extreme weather events. Under this scenario, climate change impacts could reduce the GDP per capita by 18% relative to the baseline by 2072. In the absence of any fiscal policy response, public debt levels could reach an unsustainable 140% of GDP.

The prerequisites for the effective implementation of this approach are (i) access to historical climate-related data and robust tools for analysis to enable CFAs to quantify macrofiscal implications; and (ii) effective institutional arrangements and coordination mechanisms to respond to climate change-related risks such as the cross-functional Interagency Council on Climate Change that coordinates the country's climate change response.

Climate-resilient fiscal planning can enable CFAs to mobilize and align public and private finance for investment in adaptation. Understanding the risks of implicit liabilities in state balance through SOEs and PPPs gives central finance agencies a holistic picture of risks.

Good practice elements and discussion on applicability:

- **Include fiscal health indicators in the scope and coverage of climate risk impact assessment.**
- **Conduct scenario modeling or stress testing.**

---

[a]   Government of Armenia. 2021. National Adaptation Plan.
[b]   IMF. 2022. *Armenia: Technical Report—Quantifying Fiscal Risks from Climate Change.* Technical Report; and M.E. Kahn et al. 2019. Long-Term Macroeconomic Effects of Climate Change: A Cross-country Analysis. *IMF Working Paper.* No. 2019/215.
Source: Asian Development Bank.

Climate Fiscal Disclosure and Integration into Fiscal Planning Frameworks

Box A6: The Philippines' Fiscal Risk Statement

The Philippines is facing high impacts of physical climate change. According to the Global Climate Risk Index 2000–2019, the Philippines ranks among the top 10 countries most affected by physical climate change. The average annual losses in that period amounted to approximately 30% of gross domestic product (GDP).[a] The Government of the Philippines reflects the fiscal risk arising from disasters triggered by natural hazards due to climate change in its annual Fiscal Risk Statement to "provide a comprehensive view of the country's exposure to fiscal risks emanating from fiscal projections and turnouts, public debt, and contingent liabilities."[b]

The Philippines's Fiscal Risk Statement 2023 provides an overview of the country's revenues and expenditures. It highlights the following:

- The damages associated with past disasters triggered by natural hazards and the resulting government expenditures.

- The required investment needed for the recovery and rehabilitation of affected regions until 2025 and the current period spending on flood management.

- The impact of chronic climate events on agriculture productivity (e.g., approximately $8 billion in reported damages due to El Niño).

- The current balance of the National Disaster Risk Reduction And Management (NDRRM) Fund and drawdowns from the fund in the fiscal year 2022.

- The Government of the Philippines' strategy and actions to reduce and manage the fiscal risks arising from disasters triggered by natural hazards due to climate change.

The prerequisites for the effective implementation of this approach are (i) the availability of reliable data on damages and fiscal impacts from recent disasters due to natural hazards (including estimates of investment needs); and (ii) the risk management strategy and actions in place to avoid adverse public reactions to disclosed risks.

The Philippines' Fiscal Risk Statement has been cited as good practice. Yet, besides the investment need for recovery, the Fiscal Risk Statement does not include a comprehensive forward-looking assessment of climate risk impacts and is limited to the impacts of physical climate risk. Also, there is no quantitative evidence of the impact of including climate-related risks in Fiscal Risk Statements.

Good practice element and discussion on applicability:

- **Integrate insights from climate-related fiscal risk assessment into other fiscal risk or climate-related disclosures.**

---

[a] Germanwatch. 2021. *Who Suffers Most from Extreme Weather Events? Weather-Related Loss Events in 2019 and 2000-2019.*
[b] Philippine Development Budget Coordination Committee 2023. *Fiscal Risk Statement 2023.*
Source: ADB.

Risk Assignment

Box A7: The Philippines' Risk Layering Approach

The Philippines is prone to disasters triggered by natural hazards, with 60% of the country's land area and 74% of the population exposed to hazards, including floods, cyclones, droughts, earthquakes, tsunamis, and landslides. Since 1990, the country has experienced 565 such disasters, resulting in 70,000 deaths and causing $23 billion in damages.[a] In 2013, Typhoon Haiyan caused $12.9 billion in damages and affected 12.2 million people, posing significant humanitarian and fiscal risk.[b]

In response, the Philippines developed a three-tiered risk management approach:

- **Local and national retention.** The Local Disaster Risk Reduction and Calamity Fund requires local governments to set aside at least 5% of the annual budget for disaster financing.[c] The National Disaster Risk Reduction and Management (NDRRM) Fund has allocations from the national budget and can be used for relief, recovery, and reconstruction. These funds are used to cover high-frequency, low-severity risks.

- **Contingent financing from development partners.** The Philippines has obtained contingent lending from international partners such as the World Bank, the Japan International Cooperation Agency, and the Asian Development Bank (ADB). For example, there is a $500 million loan with ADB contingent on the occurrence of disasters triggered by natural hazards or public health emergencies.[d] These funds are used to cover mid-frequency, mid-severity risks.

- **Risk transfer through traditional insurance.** Republic Act No. 656, known as the Public Insurance Law (RA 656), mandates all national government agencies, state-owned enterprises, and local government units up to first class municipalities to insure with the Government Service Insurance System, a state-owned insurance company, assets in which the government has an insurable interest.[e]

- **Risk transfer through parametric insurance.** For low-frequency, high-severity risks, the Philippines obtained parametric insurance of $389 million (as of 2018/2019) from the World Bank for national government assets against earthquakes and severe typhoons, and for 25 provinces against severe typhoons.[f]

The prerequisites for the effective implementation of this approach are (i) a robust understanding of potential losses from disasters trigged by natural hazards due to climate change to guide the design and sizing of local and national risk retention funds; (ii) the availability of resources at the local government level is crucial to support local risk retention efforts; (iii) clear cost-sharing and disbursement rules to ensure rapid disbursement and correct utilization of funds; and (iv) access to the (re)insurance market, potentially with technical assistance from multilateral development banks and international financial institutions.

continued on next page

Box A7 continued

The parametric insurance policy with the World Bank has been successful in providing rapid liquidity to the Philippines following disasters, but there have been challenges in ensuring funds reach intended beneficiaries. In 2 years, the parametric insurance programmed approximately $28 million in payouts, all within the contractually agreed timelines. The policy received strong interest from reinsurers looking to diversify their risk portfolios and the number of counterparties doubled in its second year.[g] However, funds intended for the provinces did not always reach intended beneficiaries. This was due, at least in part, to the listing of the national government as beneficiary and inter-agency procedural issues in the release of funds.

A recent technical report by the World Bank on disaster response and rehabilitation in the Philippines highlighted some potential areas for improvement in relation to the NDRRM Fund. The report highlighted that while the NDRRM Fund is a key government fund for responding to disasters, it has not reached its potential due to delays in funding release. Moreover, it has mainly been used for disaster response and provided limited support for risk reduction activities and local government units.[h]

Good practice elements and discussion on applicability:

- **Align risk assignment with required capacity:** Risks are divided into categories, with low-magnitude risks retained by local authorities that have limited capacity compared to the central government.

- **Select appropriate risk transfer instruments depending on frequency and severity of risks:** The risk layering approach ensures that cheaper sources of funding (e.g., fiscal reserve) are used for high-frequency risks, with the most expensive instruments only being used in exceptional circumstances.

---

[a] ADB and World Bank. 2021. *Climate Risk Country Profile: Philippines.*
[b] World Bank. 2014. *Recovery and Reconstruction Planning in the Aftermath of Typhoon Haiyan.*
[c] IMF. 2018. *How to Manage the Fiscal Costs of Natural Disasters.*
[d] ADB. 2019. *ADB Introduces Contingent Disaster Financing for Natural Disasters.* News Release. August 1.
[e] D. T. Villacin. 2017. A Review of Philippine Government Disaster Financing for Recovery and Reconstruction. *Philippine Institute for Development Studies Discussion Paper Series.* No. 2017-21.
[f] World Bank. 2018. *Case Study: Insuring the Philippines against Natural Disasters.*
[g] B. L. Signer and R. Poulter. 2021. Disaster Risk Insurance: 5 Insights from the Philippines. World Bank Blogs. 1 December.
[h] World Bank. 2020. Public Expenditure Review: Disaster Response and Rehabilitation in the Philippines. *Technical Report* No. 156154.
Source: ADB.

Box A8: Viet Nam Oil and Gas SOE Subsidiary Petrovietnam and Its Subsidiary, Petrovietnam Technical Services Corporation, to Support Offshore Wind Development

Viet Nam faces material transition risks, with a sizeable share of gross domestic product (GDP) arising from highly exposed sectors such as high-emission manufacturing, fossil-fuel-based power, and agriculture.[a] A significant share of Viet Nam's total energy supply in 2021 came from coal (49.1%) and oil and gas (30.7%).[b]

There is a need for state-owned enterprises (SOEs) in oil and gas to manage the transition. Where the need can be justified in terms of wider economic interests (e.g., to bring technologies to maturity), this may involve making strategic investments in green technologies such as offshore wind. Since 2022, Viet Nam's state oil and gas company, Petrovietnam, with support from the Asian Development Bank (ADB), has been preparing a Green Energy Transition Strategy, Roadmap, and Action Plan. In 2022–2023, a pilot carbon capture and storage project with a capacity of 50,000 tons per year at the Tien Hai depleted onshore field was proposed for Petrovietnam to invest in. In 2023, Petrovietnam updated the power targets at an installed capacity of 8-14 gigawatts including 5%–10% by renewable energy (RE) by 2030, and an installed capacity of 8%-10% of the total national installed capacity including 10%–20% by RE by 2030. Petrovietnam is also planning develop offshore wind power projects for green hydrogen.[c]

Offshore wind presents a significant opportunity for energy players, but domestic developers have highlighted the need for technical support from experienced international offshore wind partners. In May 2023, Viet Nam's Prime Minister approved the eighth power development plan that includes ambitious offshore wind target (6,000 megawatts [MW] by 2030 and 70,000–91,500 MW by 2050). Currently, there is no operational offshore wind project.[d]

To kickstart the development of offshore wind, initial pilot projects will be carried out by SOEs (Vietnam Electricity and Petrovietnam) or the Ministry of Industry and Trade. The assignment of foreign and private investors will be considered after a comprehensive evaluation of the pilot projects.[e]

Petrovietnam Technical Services Corporation (PTSC) is a subsidiary of Petrovietnam. Its core businesses initially involved providing technical services for the oil and gas industry. Given its technical capabilities and experience in offshore, PTSC has the potential to pioneer the development of offshore wind power projects in Viet Nam.[f]

- In 2021, following approval from Petrovietnam and the resolution of the general meeting of the shareholders, PTSC was authorized as the developer or investor of offshore wind projects, making it Petrovietnam's only subsidiary with full legal grounds to invest in offshore wind.

- In 2023, PTSC received authorization from the Ministry of Natural Resources and Environment to conduct wind, marine, and geological surveys for an offshore wind farm project in the waters off the Ba Ria – Vung Tau province.

continued on next page

Box A8 continued

- PTSC and Sembcorp Utilities in Singapore signed a joint development agreement for investment in offshore renewables in Viet Nam and the export of green electricity to Singapore.

The prerequisites for the effective implementation of this approach are (i) stakeholder buy-in, including support from owning line ministries; (ii) complementary policy support from line ministries to ensure a coherent and supportive regulatory environment; and (iii) technical capabilities or transferable expertise. Establishing a subsidiary with greater flexibility could facilitate this transition.

Good practice element and discussion on applicability:

- **Promote risk management practices through identifying new, greener business opportunities for SOEs in transition:** PTSC leverages its technical expertise in offshore oil and gas to advance the development of pilot projects; as an SOE, it plays a crucial role in advancing the development of offshore wind policy frameworks (e.g., leases and licenses) before foreign and private investors are assigned (footnote d).

---

[a] McKinsey & Company. 2022. *How the Net-Zero Transition Would Play out in Countries and Regions.*
[b] IEA. Countries & Regions. Asia-Pacific. Viet Nam.
[c] Petrovietnam. 2024. *Petrovietnam tiên phong trong chiến lược năng lượng xanh.* 21.
[d] A. Urakami. 2024. *Preliminary Evaluation of Obstacles and Policy Measures to Facilitate Offshore Wind Power Expansion in Vietnam.* Discussion Paper. No. 70. Kyoto University.
[e] C. T. Duc. 2024. Challenges and Solutions for the Pilot Project Mechanism Aims to Develop Policies for Offshore Wind Power in Vietnam. LinkedIn. 29 April.
[f] Petrovietnam Technical Services Corporation (PTSC). Services- Renewable Energy; and PTSC. 2023. PTSC Granted the Approval for Conducting Comprehensive Surveys for an Offshore Wind Farm Project in Vietnam, Aimed at Facilitating the Export of Electricity to Singapore. 29 August.
Source: ADB.

Climate Investment and Financing Strategy

Box A9: Cambodia Assessed Investment Needs for the LTS for Carbon Neutrality

To achieve the net-zero transition, it is crucial to develop long-term investment strategies that clearly define the investment needs, trajectory, and resources available to support the investment. However, investment planning for the transition beyond 2030 is limited for most countries. This absence of a long-term view on the investment trajectory poses challenges to redirecting investment at scale.

Cambodia has advanced in this area, assessing the total public financing required to support its long-term strategy (LTS) for carbon neutrality. For 2025–2050, the total financing needs were estimated at $9 billion. A year-on-year (2022–2050) financing needs trajectory is also provided.[a]

A high-level financing plan was also developed to provide clarity on potential sources of funding to support envisioned investments. The proposed financing plan involves allocating 1% of new public borrowing toward climate-related initiatives and redirecting 2% of public spending on economic services. The plan also includes implementing a pricing policy and taxation reforms in the transportation sector with the aim of covering 90% of the investment needs in this sector. Lastly, international climate finance will be leveraged to fund the remaining 26% of the total requirement (footnote a).

The LTS also discussed the required private investment to support Cambodia's transition to carbon neutrality. Estimates indicate an increase from $0.5 billion annually in 2030 to $1.4 billion by 2050. Key sectors identified for private investment include energy, transportation, and forestry (footnote a).

The prerequisites for the effective implementation of this approach are (i) an aligned national vision on the transition pathway or adaptation target, supported by a concrete set of actions to inform the needs assessment; (ii) technical capacity, robust methodologies, and reliable data to support cost estimates—this could be particularly challenging for adaptation measures, given the limited availability of data and costing methodologies; and (iii) a clear understanding of the risk–return appetite of different types of capital and the potential barriers to private investment.

Good practice elements and discussion on applicability:

- **Estimate and publicize long-term investment needs:** Cambodia's LTS outlined financing needs for 2022–2050.
- **Explore feasible financing splits:** Sectors suitable for public versus private investment were identified.

---

[a] Government of Cambodia. 2021. *Long-Term Strategy for Carbon Neutrality.*
Source: ADB.

Box A10: The Philippines Introduced Climate Budgeting Reforms

The Climate Change Act of 2009 mandates the Philippines' Department of Budget and Management (DBM) to formulate the national budget in a way that ensures the appropriate prioritization and allocation of funds to support climate change-related programs. Several budgeting reforms have been implemented, with technical assistance from the World Bank in developing, implementing, and institutionalizing them.

Climate budgeting reforms encompass multiple stages of the budget cycle.[a]

- **Budget preparation.** A dedicated budget program, Climate Change Adaptation and Risk Resilience Program, was established to ensure the availability of a dedicated finances for adaptation. The activities under this program are aligned with the climate change adaptation and risk resilience plan.

- **Budget approval.** Budget briefs were developed to identify the scope and scale of sector agencies' climate budgets, gaps in coverage, and activity type; three iterations are made before execution.

- **Execution, monitoring, and reporting.** Climate change expenditure tagging was introduced in which climate budget reports are developed based on tagged spending.

The prerequisites for the effective implementation of this approach are (i) strong political buy-in to drive and coordinate reform efforts, (ii) the institutional capacity of central finance agencies and budget holders, and (iii) a clear and simple green budget taxonomy to identify and track green expenditures. Technical assistance from international financial institutions or multilateral development banks in supporting initial design could be valuable.

The national budget allocation for climate actions experienced a compound growth rate of 22% from 2016–2019. The Philippines has developed an enhanced capacity to make climate-informed budgetary decisions. As the technical assistance from the World Bank was scaled back, the use of the tagging system has been sustained (footnote a).

However, challenges also emerged in the process of climate-tagging, due to a limited understanding of the concept of "climate adaptation," especially among local authorities. This has led to inconsistencies in classifying funding budgeted for climate-related purposes (footnote a).

Good practice elements and discussion on applicability:

- **Leverage initial strategic phase of the budget formulation process:** Climate budgeting reforms have been integrated into official circulars and guidelines.
- **Establish clear public accounting standards and unified criteria:**
    » In 2012, climate change expenditure tagging was introduced. Spending lines are either marked as climate change-related or not; this simple design facilitated its rapid rollout.
    » In 2024, the DBM and the Climate Change Commission hosted a Climate Change Expenditure Tagging Orientation for all departments, agencies, and state-owned enterprises, to provide guidelines on the tagging of government expenditure on climate change.[b]

---

[a] S. Allan et al. 2019. The Role of Domestic Budget in Financing Climate Change Adaptation: A Background Paper for the Global Commission on Adaptation. 23 December.
[b] Government of the Philippines, Department of Budget and Management. 2024. Circular Letter No. 2024-6. 26 February.
Source: ADB.

Box A11: The Philippines Integrated Climate Considerations into the PPP Evaluation

Public–private partnership (PPP) models could be leveraged to secure private sector capital and capabilities. Climate-related risk considerations may be integrated into PPP arrangements to unlock their potential benefits while minimizing risks to central financial agencies (CFAs).

In 2009, the Philippines adopted a Climate Change Act that required climate issues to be considered when planning new public investments. It has since taken steps to integrate climate risk considerations into PPP contract evaluation.

In 2018, the PPP Governing Board published a resolution on including environmental concerns in PPPs, covering (i) safeguards from the impact on the environment; (ii) resilience to natural hazards, including climate change hazards; and (iii) climate change mitigation.[a] Concrete actions listed for implementing agencies include the following:[b]

- assessing natural hazard vulnerability in the selection of project locations;

- identifying applicable climate and natural hazards;

- incorporating adaptation and risk management into project design; and

- preparing adaptation and mitigation strategies, such as energy conservation measures.

The prerequisites for the effective implementation of this approach are (i) implementing agencies possess the technical capacity to conduct comprehensive climate risk assessments and develop effective mitigation strategies and (ii) CFAs have dedicated project appraisal capacity regarding climate mitigation and adaptation across key sectors relevant to PPPs.

PPP has been successfully adopted in various local mitigation and adaptation projects, such as the Quezon City waste–to–energy project, Baguio water supply project, Puerto Galera sewerage and wastewater treatment plant project, Manila Bay integrated flood control and coastal defense project, Angat hydroelectric power plant project, etc.[c]

Good practice element and discussion on applicability:

- **Establish clear guidelines to embed climate considerations from the project identification and design stage:** The Philippines integrates climate resiliency, adaptation, and mitigation considerations into the design of potential PPP projects.

---

[a] Government of the Philippines, PPP Governing Board. 2018. *Resolution No. 2018-12-02 Safeguards in PPP: Mainstreaming Environmental, Displacement, Social and Gender Concerns.* 14 December.
[b] World Bank. 2022. Reference Guide for Climate-Smart Public Investment.
[c] Government of the Philippines, PPP Center. 2019. Climate-Resilient Infrastructure Protecting the Filipinos' Future. PPP Talk: January–June 2019.
Source: ADB.

Box A12: The Malaysian Green Public Procurement Initiative

Green public procurement (GPP) is a powerful tool to use public sector purchasing to support climate goals; the impact is especially significant in markets with a large share of public procurement (e.g., buildings, construction, and public transportation).

The implementation of GPP began in 2013 with an initial short-term action plan for 2013–2015 involving five ministries. After the successful implementation of this pilot phase, the government developed a long-term action plan for 2016–2030, which progressively expanded GPP coverage to 12 ministries and their agencies in 2016 and extended to all 25 ministries and their agencies in 2017.[a]

After achieving the target of 20% GPP by 2020, a new target of 25% by 2025 was introduced, and the scope of the initiative is now being expanded to include state and local authorities, with pilot projects being implemented in three state governments and two local authorities.[b] Capability-building initiatives were also launched to support the integration of GPP into procurement processes. The pilot projects trained over 400 implementers, suppliers, and contractors to support the successful implementation (footnote b).

The prerequisites for the effective implementation of this approach are (i) institutional capacity and a robust data infrastructure for accurate data collection and reporting to ensure transparency and accountability; (ii) the availability of nationally recognized ecolabels or certification schemes to provide a basis for product and service selection; and (iii) knowledge about green products and services, and their availability in the market.

From 2013–2020, the total GPP value amounted to RM1,754 million ($374 million), and 16,519 tons of carbon dioxide (CO_2) equivalent/year emission reduction was achieved.[c]

Good practice elements and discussion on applicability:

- **Establish a transparent monitoring and evaluation framework:[d]**
  - » Monitoring data is collected annually, and the Ministry of Finance is upgrading its e-procurement system to streamline GPP data collection and tracking.
  - » The GPP system monitors four impact indicators (i.e., energy savings, energy cost savings, economic savings, and CO_2 emission reduction).
  - » Biannual reports on the implementation of GPP will be submitted to the Ministry of Finance and the Malaysian Green Technology and Climate Change Corporation; results on GPP performance and total emission reduction achieved will be presented at GPP steering committee meetings.
- **Use simple and standardized green criteria:** GPP criteria were established based on various national and international ecolabeling schemes, for example, the Malaysia Type I ecolabel and the energy and water conservation schemes (footnote a).
- **Utilize detailed market research to keep the GPP criteria up to date:** The GPP criteria were adapted wherever necessary to ensure adequate product availability (footnote a)

[a] Switch Asia. 2020. *Monitoring the Implementation and Estimating the Benefits of Sustainable/Green Public Procurement.*
[b] GIZ. 2024. *Malaysia Expands Government Green Procurement (GGP) to State and Local Authorities.* 1 April.
[c] Switch Asia. 2024. *Malaysia Country Report: Enhancing Green Public Procurement implementation.*
[d] International Institute of Sustainable Development. 2024. Monitoring Progress in Green Public Procurement: Methods, Challenges, and Case Studies.
Source: ADB.

Ensuring Sustainable Fiscal Trajectory

Box A13: Indonesia Fiscal Policy Agency Study for LPG Subsidy Reform

The Government of Indonesia is considering reforming the liquefied petroleum gas (LPG) subsidy to improve focus on low-income households and exposure of the fiscal budget to gas price volatilities.[a] The LPG subsidy introduced in 2008 set the LPG price at Rp4,250 per kilogram (kg). The subsidized consumer price supported the 93% decrease in kerosene consumption for cooking between 2008 and 2016.

The design had two shortcomings:

- The subsidy was regressive, benefiting higher-income households more.
- The subsidy depended on increasing external LPG prices, which strained the fiscal budget (approximately 2.5% of total state expenditures in 2020).

The Ministry of Finance's fiscal policy agency and PROSPERA conducted a study to develop a reform proposal supported by a socioeconomic impact assessment (e.g., impact on inflation, poverty rate, and inequality indicators) and an assessment of the expected fiscal burden. The study was enabled by the wealth of robust and comprehensive data on households' income, other socioeconomic indicators (e.g., household size), and average LPG consumption.

To ensure a progressive fiscal benefit, eligible low-income households were granted a direct cash transfer for each 3 kg cylinder of LPG purchased or for a predetermined number of cylinders per month.

The prerequisites for the effective implementation of this approach are (i) the availability of an accessible register of eligible households (e.g., the existing registry on energy subsidy eligibility); (ii) the development of infrastructure to failproof registered LPG consumption (e.g., through the use of biometric data), and (iii) the deployment of infrastructure to process and disburse subsidy payments.

Good practice elements and discussion on applicability:

- **Conduct robust incidence assessments.**
- **Target transition support mechanism.**

---

[a]   J. Kuehl et al. 2021. *LPG Subsidy Reform in Indonesia: Lessons Learned from International Experience*. Global Subsidies Initiative Report. International Institute for Sustainable Development.

Source: ADB.

Box A14: Indonesia's Fossil Fuel Subsidy Phase-Out

In 2000, fossil fuel subsidies reached around 20% of the Government of Indonesia's budget, constraining public resources. The government has since taken various steps to reform fossil fuel subsidies, which are often accompanied by measures to mitigate the negative impact on low-income communities, including the following:[a]

- **Prior to 2005, prioritized subsidy reform for gasoline and diesel over kerosene.** Gasoline and diesel are used by car and motorcycle owners, whereas kerosene was the primary energy source among the urban poor. The 2005 fuel subsidy reform led to a significant increase in kerosene price, and the government introduced an unconditional cash transfer program to mitigate the impact of the reform on poor people.

- **Prioritized removal of subsidies for industries.** Since 2005, industries are no longer eligible for subsidized diesel.

- **Mitigation measures for the 2005 gasoline and diesel price reform.** Cash transfers to low-income households (BLT program); reallocation of savings to education (School Operational Assistance program), health (Health Insurance for the Poor program), and infrastructure programs (Rural Infrastructure Support Project).

- **Mitigation measures for the 2007 kerosene–liquefied petroleum gas (LPG) conversion program.** Increased subsidies for liquefied petroleum gas (LPG)—a cleaner and cheaper-to-subsidize alternative cooking fuel—and the provision of free LPG starter packages to households and microbusinesses.

- **Collaboration with line ministries and provincial governments to ensure targeted support for affected communities.** A task force was set up for the LPG conversion program, involving the National Team for Poverty Alleviation and the Ministry of Social Affairs. Provincial governments also played a key role in targeting low-income households for conversion.

- **Leveraged technology to ensure targeted support.** Smartcard systems for health, education, and family spending were implemented to mitigate the problem of frivolous spending and corruption associated with direct cash transfers.

The prerequisites for the effective implementation of this approach are (i) availability and affordability of cleaner alternatives for impacted industries and households for a smooth phase-out of fossil fuel subsidies; (ii) availability of an accessible register of eligible households and vehicle types (e.g., existing registry on energy subsidy eligibility), households' access to personal bank accounts, and development of infrastructure to process and disburse subsidy payments to eligible households; and (iii) ability to reallocate subsidy savings (e.g., ring-fenced pool).

Good practice elements and discussion on applicability:

- **Apply a phased approach.**
- **Target transition support mechanism.**
- **Conduct stakeholder engagement and supporting communication campaigns.**

---

[a] International Institute for Sustainable Development. 2010. *Lessons Learned from Lessons Learned from Indonesia's Attempts to Reform Fossil-Fuel Subsidies*; and R.K. Gobel et al. 2014. Equity and Efficiency: An Examination of Indonesia's Energy Subsidy Policy and Pathways to Inclusive Reform. *MDPI Sustainability.* 16 (1): 407.
Source: ADB.

Box A15: Singapore's Carbon Tax

In 2019, Singapore introduced a carbon tax. The price was initially set at S$5 per ton of carbon dioxide equivalent (tCO_2e) for 2019–2023 and was raised to S$25/$tCO_2e$ in 2024. It will be raised to S$45/$tCO_2e$ in 2026–2027, with a view of reaching S$50–S$80/tCO_2e by 2030. According to Singapore's Reach Government Feedback Unit, a S$10–S$20/tCO_2e tax is similar to an increase in current electricity prices by 2.1% to 4.3%.

Key success factors included the following:

- **Phase-in of the tax through a gradual increase in CO_2 tax and a sector approach.** The carbon tax is to increase gradually with the schedule being clearly communicated in advance. The carbon tax is applied on a sectoral basis, with an initial focus on power generation and petrochemical industries that contribute significantly to emissions.

- **Adjustable and calibrated transition allowance framework to incentivize investment in decarbonization.** From 2024, a transition framework will be introduced to give emissions-intensive trade-exposed companies more time to adjust to a low-carbon economy. Transition allowances will be based on internationally recognized efficiency benchmarks or companies' decarbonization plans. These allowances will be reviewed and adjusted considering companies' decarbonization performance and advancements in decarbonization technologies.
    - » **Grants to support energy efficiency.** The government has launched energy efficiency programs to support businesses and households.
    - » **Energy efficiency fund.** Funds 70% of the costs incurred in the adoption of pre-approved energy-efficient technologies by manufacturing companies.

- **Climate vouchers.** All Housing and Development Board households can claim S$300 for the purchase of energy- and water-efficient appliances.

- **Stakeholder management and consultations.** In 2017, a public consultation was launched on the carbon pricing bill draft. Responses addressing concerns raised during the consultation (e.g., minor emissions with high monitoring costs) were then published. Additionally, although carbon tax revenues are not ring-fenced, the government has communicated that it intends to spend more than this amount on projects that deliver emissions abatement.

- **Use of international carbon credits.** Starting 2024, companies may use high quality international carbon credits, in compliance with rules under Article 6 of the Paris Agreement, to offset up to 5% of their taxable emissions.

The prerequisites for the effective implementation of this approach are (i) ex-ante impact assessments on sectors and the economy, (ii) technical capability to calibrate the transition allowance based on the technological evolution, and (iii) a robust monitoring and verification framework.

Good practice elements and discussion on applicability:

- **Apply a phased approach.**
- **Target transition support mechanism.**
- **Conduct stakeholder engagement and supporting communication campaigns.**

Source: ADB.

Box A16: Malaysia's Solar Industry Ramp-Up

The Government of Malaysia aligned fiscal incentives, and energy sector and investment policies to support the solar photovoltaic (PV) industry with a focus on developing capital-intensive cell and module production.

- **Tax incentives.** Income tax exemptions and investment tax allowances were introduced for the purchase of green technology assets for the solar industry.

- **Soft loans.** The Green Technology Financing Scheme (GTFS) provided loans to producers and users of green technology with a 2% interest subsidy from the government and a 60% government guarantee.

- **Energy policies.** The feed-in tariff allowed electricity from renewables to be sold to utilities at a fixed premium (ended in 2017 due to limited funding); replaced by net energy metering which allowed excess electricity generated by solar PV installed for own use to be exported back to the grid in exchange for cash.

- **Investment policies.** The Large-Scale Solar Scheme awards contracts to build solar power plants in the range of 1–50 megawatts (MW), with a 1,250 MW total target installation capacity.

Many of these policies were launched by the Ministry of Finance as part of a broader budget setting process. Furthermore, the national strategy to support the solar PV sector is embedded in a broader National Energy Transition Roadmap, which has been designed with an explicit objective of ensuring fiscal sustainability.

In part as a result of these measures, Malaysia has become a solar PV production center, attracting investment from multinationals. As of 2015, the Malaysian Investment Development Authority is estimated to have attracted $7.2 billion investment in solar component manufacture (of which 95% is foreign investment).[a] Solar capacity increased between 2015 and 2020 from 273 gigawatt-hours (GWh) to 2,044 GWh, as did and renewables penetration, doubling from 3% to 6%.[b] Malaysia's Performance Management and Delivery Unit estimated an increase in gross national income from solar power to $112 million by 2020 as a result of supportive government policies (footnote a).

The prerequisites for the effective implementation of this approach are (i) collaboration across ministries, (ii) sufficient funding for continued implementation of policies, and (iii) concurrent reduction on fossil fuel subsidies.

Good practice elements and discussion on applicability:

- **Use of market mechanism and efficient risk sharing.**
- **Close coordination of fiscal incentives.**
- **Explore expected impact on government revenues.**

---

[a] U. E. Hansen and I. Nygaard, eds. 2019. *Trade in Environmentally Sound Technologies in the ASEAN Region.*
[b] IEA. Countries & Regions. Asia Pacific. *Malaysia.*
Source: ADB.

Box A17: Thailand Electric Vehicle Industry Expansion

The automotive industry plays a significant role in Thailand's economy, contributing approximately 10% to its gross domestic product (GDP). As the 11th largest automobile production hub, Thailand has an end-to-end production ecosystem and the associated technical know-how and labor force. The transition to electric vehicles (EVs) presents challenges and great opportunities for the local automobile industry. The government has prioritized the development of "next-generation automotive" as a key industry, with the strategic aim of transitioning from an internal combustion engine vehicle hub to an EV production hub in Southeast Asia.

The National Electric Vehicles Policy Committee (including the Ministry of Finance) developed a three-phase road map for EV development from 2021–2035. The road map covers not only EV production and usage, but also battery manufacturing and supplies, charging stations, power grid management, and the development of safety standards and regulations.

Many of the specific incentives underpinning EV policies were launched by the Ministry of Finance as part of a wider budget process. These included measures set out in the "EV Stimulation Package (EV3.0)," which ranged from vehicle subsidies to reductions in excise taxes and import tariffs.[a] Incentives were adjusted over time to reflect level of need and fiscal affordability.[b] However, these processes could be made more effective with greater transparency on the level and efficacy of expenditures.[c]

The prerequisites for the effective implementation of this approach are (i) political buy-in across line ministries, with clear and aligned policy objectives and a long-term road map to guide actions; (ii) close engagement with industry and investors; (iii) integrator or facilitator role played by Board of Investment to integrate government support tools into a coherent package and coordinating with various agencies to develop the EV ecosystem; and (iv) technical capacity in incentive design in line with market and technology development.

Good practice elements and discussion on applicability:

- **Shape sector transition road maps.**
- **Close coordination of fiscal incentives.**
- **Appropriate incentive time frame.**
- **Monitor fiscal affordability.**

---

[a] Government of Thailand, Board of Investment. 2023. *Opportunities and Support Measures for EV Activities*. Policy presentation.
[b] *S. Strangio*. 2023. Thailand Announces Reduced Subsidies for EVs as Sales Boom. *The Diplomat*. 2 November.
[c] United Nations Development Programme. 2023. Development Finance Assessment for Thailand. April.
Source: ADB.

Building Up Fiscal Resilience Against Climate-Related Shocks

Box A18: The Southeast Asia Disaster Risk Insurance Facility Risk Pool

The Lao People's Democratic Republic (the Lao PDR) is exposed to a range of climate-related hazards, including droughts, floods, and storms. Even though its geographic location shields the Lao PDR, past disasters significantly impacted gross domestic product (GDP); for example, the 2018 flood decreased GDP by 2%.[a] Additionally, the Lao PDR has the fourth lowest nonlife insurance penetration and limited or no insurance for private homes, crops, livestock, or public assets. The limited insurance access has created significant liabilities for the Government of the Lao PDR: the post-disaster funding need in 2018 was 10% of the Lao PDR's annual budget, and there is an estimated funding gap exceeding 300% for 1-in-30-year events.

In 2018, Cambodia, Indonesia, Japan, the Lao PDR, Myanmar, and Singapore agreed to establish the Southeast Asia Disaster Risk Insurance Facility (SEADRIF) as a regional platform for Southeast Asian countries to access disaster risk financing solutions and increase financial resilience to climate and disaster risks. Following incorporation in Singapore in 2019, the SEADRIF developed a system for assessing the level of flood risks and monitoring live flood events in the Lao PDR and Myanmar. In parallel, the SEADRIF worked with the re/insurance market to develop options for structuring a catastrophe risk insurance product.[b]

In 2021, SEADRIF launched its first regional catastrophe risk insurance product for the Lao PDR against flood risks (with ongoing discussions with other countries), and the Lao PDR, with financial support from the World Bank, paid $5 million premium for subscription. The insurance includes two complementary components:[c]

- A parametric component that automatically triggers payouts if a flood event exceeds a specified threshold. The parametric component is paid out fast, latest within 10 days.

- A finite component that covers damages of events that do not trigger the parametric component or damages that exceed the parametric payout. The component is paid out 5 days after the Lao PDR submits a notice of loss. The payout is commensurate to the actual losses, but the payout time is longer.

The Lao PDR also committed to developing contingency plans to ensure timely, effective, and transparent use of the insurance payouts (e.g., use to be reported within 9 months). Payouts would be used for post-disaster relief, recovery of critical infrastructure, or maintenance of essential government services.

The prerequisites for the effective implementation of this approach are (i) a country-specific flood risk model to support insurance pricing; (ii) historic loss data instead of an asset database; (iii) a tool to monitor the severity of floods and determine insurance triggers; (iv) collaboration with neighbor countries, multilateral development banks, and donors to create risk pool; (v) an independent company or agency with audit responsibility; and (vi) significant support from international donors.

continued on next page

Box A18 continued

Faced with extensive flooding across 10 provinces in August 2023, the Lao PDR submitted a notice of loss forms and received two payments totaling $1.5 million from SEADRIF in support of flood response and rehabilitation within just one business day. In February 2024, the Lao PDR renewed the parametric flood insurance policy, with revisions to expand the parametric trigger from two levels to four levels and to lower thresholds, providing flexibility to address varying risk scenarios.[d] After reviewing provisional data on the number of people affected by the floods across eight provinces in September 2024, SEADRIF initiated a first payout of $0.75 million within a week and announced that an additional payout may be triggered as updated data on Mekong River water levels becomes available.

Good practice elements and discussion on applicability:

- **Set up a system of financing mechanisms.**
- **Pair financing mechanism with well-designed triggers and management rules.**
- **Develop or pilot solutions in partnership with ASEAN members, MDBs, IFIs, and private sector stakeholders.**

---

[a]   United Nations Office for Disaster Risk Reduction. 2019. *Disaster Risk Reduction in Lao PDR: Status Report 2019*.
[b]   JBA Risk Management. 2024. *Building Financial Resilience to Flood Risk in South-East Asia*.
[c]   World Bank. 2020. *Southeast Asia Disaster Risk Insurance Facility (SEADRIF): Strengthening Financial Resilience in Southeast Asia*. Project Appraisal Document.
[d]   SEADRIF Insurance Company. Frequently Asked Questions.
Source: ADB.

Box A19: Indonesia's Pooling Fund for Disasters

Indonesia faced significant budget spending on disaster response and recovery. Between 2014 and 2018, the Government of Indonesia spent $90 million to $500 million annually on disaster response and recovery, which is equivalent to approximately 2% of the central government's budget. The subnational governments spent an additional estimated $250 million. Indonesia recognized the risk to the budget and indirectly to the government's ability to finance other strategic priorities. In response, the government established the pooling fund for disasters (PFB) to increase its response capacity without reallocating budgetary resources.[a]

Indonesia established an innovative central disaster pooling fund:[b]

- Indonesia is setting up a fund to manage disaster finance, to be hosted by the Indonesian Environment Fund.[c] The pooling fund will be used to invest in insurance as well as capital markets. The insurance payouts and returns are intended to cover future disaster relief financing needs.

- The PFB centralizes Indonesia's national and subnational contingency funds and purchasing of risk transfer products. National and subnational contingency budgets will not be replaced completely but are being complemented.

- The central mechanism is supposed to help ensure an effective and transparent flow of money to relevant government agencies, including faster social assistance payments for victims of disasters and improve preparedness planning.

- To ensure effectiveness, the PFB will also invest in activities to improve planning, such as introducing budget tracking for disaster-related expenditures.

- The program aims to create strong ownership across different government agencies and develop technical capacity through knowledge-sharing sessions for the government officials with international partners (e.g., a workshop in Switzerland on Swiss Disaster Risk Finance and Management strategy).

The prerequisites for the effective implementation of this approach are (i) good access to financial markets (including the private insurance market), (ii) technical capabilities to manage funds, (iii) buy-in from national and subnational stakeholders, (iv) access to relevant local data as the basis for planning and response coordination, and (v) and significant support from international partners.

The initiative resulted in the allocation of approximately $500 million in government funds to the PFB, based on a loan provided by the World Bank and a $14 million grant received from the Global Risk Financing Facility to support its operationalization. However, PFB is still being developed, so additional impact is yet to be achieved.

Good practice elements and discussion on applicability:

- **Set up a system of financing mechanisms.**
- **Create a central team to assess, monitor, and coordinate risk management across different government layers.**
- **Proactively engage stakeholder engagement and share knowledge.**

---

[a] World Bank. 2021. *How Indonesia Strengthened its Disaster Response with Risk Finance and Insurance*. 17 November.
[b] World Bank. 2023. *Sovereign Disaster Risk Finance and Insurance in Middle-Income Countries Program Review (2017-2022)*.
[c] Global Shield Financing Facility. 2024. *Indonesia Knowledge Exchange Visit*. January.
Source: ADB.

Appendix III. Case Studies: Climate Finance Mobilization

Mobilize and Deploy Public Funds

Box A20: Indonesia's Green Bond and Green Sukuk Initiatives

Indonesia developed the green bond and green *sukuk* (Islamic bond) to finance for green projects. In 2015, the Government of Indonesia introduced the Budget Tagging Mechanism; in 2018, it was expanded to cover climate change mitigation and adaptation, involving 17 line ministries. In 2018, the Ministry of Finance, National Development Planning Agency, and Ministry of Environment collaboratively supported the issuance of the world's first green *sukuk*.[a] This initiative would finance "tagged" eligible green projects for transition and adaptation, appeal to climate-conscious capital, and sometimes secure a green premium. Indonesia developed a following framework for green bond and green *sukuk*:

- **Use of proceeds.** Eligible green projects in nine sectors (renewable energy, energy efficiency, resilience to climate change, green buildings, etc.).

- **Project evaluation and selection.** Review and approval by the Ministry of Finance and National Development Planning Agency.

- **Management of proceeds.** The Ministry of Finance manages the allocation of proceeds, which will be credited to a designated account of relevant ministries for funding predefined projects.

- **Reporting.** The Ministry of Finance will prepare and publish an annual green bond and green *sukuk* report, based on information reported by line ministries utilizing the proceeds

The prerequisites for the effective implementation of this approach are (i) a pipeline of high-quality green investment projects; (ii) transparent monitoring, evaluation, and reporting mechanisms; (iii) investor appetite for green bonds issued by the sovereign; (iv) capacity and expertise to manage issuance processes; and (v) large ticket sizes to absorb additional issuance costs associated with "greenness" verification.

The green *sukuk* and green bond initiatives created a significant impact:[b]

- **Four issuances in 2018–2020, totaling $3.2 billion.** Government issuances have gained positive responses from investors. The green *sukuk* issued in February 2019 was oversubscribed by 3.8 times; issuance in June 2020 was oversubscribed by 7.4 times.[c]

- **Visible green premium offering a borrowing cost advantage.** In June 2021, Indonesia priced its fourth green *sukuk* and the green yield curve remained inside the vanilla curve.

Good practice element and discussion on applicability:

- **Leverage green or sustainability-linked bonds.**

---

[a] United Nations Development Programme. 2018. Indonesia's Green Bond & Sukuk Initiative.

[b] Asian Development Bank (ADB). 2022. *Green Bond Market Survey for Indonesia: Insights on the Perspectives of Institutional Investors and Underwriters.*

[c] R. F. Prisandy and W. Widyaningrum. 2022. Green Bond in Indonesia: The Challenges and Opportunities. In Y. B. Wang, R. Sagita, and D. Sugiharti, eds. *Indonesia Post-pandemic Outlook: Rethinking Health and Economics Post-COVID-19.* pp. 259–278. BRIN Publishing.

Source: ADB.

Box A21: Indonesia's Regional Infrastructure Development Fund

Investors have highlighted insufficient infrastructure as a significant barrier to investment in Indonesia. A key contributing factor is the limited availability of financing options for subnational governments, which are responsible for about 60% of public infrastructure investment. Moreover, subnational governments often have limited technical and institutional capacity to ensure effective investments in strategic infrastructure.

Indonesia established the Regional Infrastructure Development Fund (RIDF) to finance subnational governments. In 2017, the Ministry of Finance (MOF), in partnership with the World Bank and the Asian Infrastructure Investment Bank (AIIB), established the RIDF under PT Sarana Multi Infrastruktur (PT SMI), the special mission vehicle under the MOF.[a] The goal was to increase access to infrastructure finance at the subnational level through a financially sustainable intermediary.

- RIDF was designed to lend directly to subnational governments so they can address their critical infrastructure needs more effectively and overcome annual funding constraints. This program could also build the credit history and borrowing experience of subnational governments.

- A complementary RIDF project development facility was set up to support subnational governments in carrying out subproject identification and preparation, including feasibility studies, detailed engineering designs, safeguards assessments, advisory services on financial management and procurement, and trainings. This technical assistance was provided in thematic areas of urban water supply, drainage, and flood risk reduction.

The prerequisites for the effective implementation of this approach are (i) a legal and regulatory framework that supports subnational borrowing, monitoring, and financial disclosure; (ii) the capacity of subnational governments to plan, finance, and implement climate projects; and (iii) technical support from international financial institutions for the design and initial implementation of the facility.

Full operationalization of the RIDF was delayed by the challenges related to the preparation of technical documents, procurement issues, the coronavirus disease (COVID-19) pandemic, and competition from emerging alternative financing sources such as the National Economic Recovery loan program. However, the RIDF had a significant impact:[b]

- **Material number of projects funded and supported.** Before RIDF, PT SMI had only seven subnational government projects (four roads and three hospitals projects) in its portfolio. After 5 years, PT SMI's portfolio had increased to 42 projects, including projects in public markets, water supply, and transportation. The financing of 12 projects came from RIDF, while the remaining 30 were financed by PT SMI's equity. The RIDF loan portfolio had an average loan size of approximately $13.1 million, and the rate of return on RIDF assets was 2%.

- **Capacity of subnational governments strengthened.** The RIDF project development facility services were provided to 30 subnational governments. A total of 89 subnational governments received capacity building trainings on a range of topics.

- **CFA institutionalized regulations to subnational government borrowing.** A security mechanism was introduced to enable pledging future fiscal transfer revenues as collateral for the loan, and was subsequently adopted for all other subnational lending instruments administered by PT SMI.

Good practice elements and discussion on applicability:

- **Pilot national-to-local loan programs.**
- **Provide technical support and institutional capacity building.**
- **Provide credit enhancement.**

---

[a] AIIB. 2017. Regional Infrastructure Development Fund: Project Summary Information.
[b] World Bank. 2024. Regional Infrastructure Development Fund: Implementation Completion Report (ICR) Review; and M. Sharma, J, Lee, and W. Streeter. 2023. Mobilizing Resources Through Municipal Bonds: Experiences form Developed and Developing Countries. *ADB Sustainable Development Working Paper Series.*

Source: ADB.

Box A22: Viet Nam's Agribank Boosts Green Lending Through Preferential Credit Programs

Climate mitigation and adaptation are critical for agriculture, forestry, and fishing, which make up a major part of the economy of Viet Nam (12% of gross domestic product [GDP] as of 2023). Agriculture is the third-highest emission sector. On the other hand, climate-related physical hazards such as high temperatures, droughts, and floods threaten the production of key food and cash crops. Agribank is a state-owned bank focused on credit for agricultural and rural development. Agribank aims to promote green credit growth to support businesses, farmers, and cooperatives; and at the same time manage environmental and social risks in credit extension activities.

In Viet Nam, Agribank incorporated environmental, social, and governance (ESG) in its business strategy to boost green credit. In 2017, Viet Nam's Prime Minister required the State Bank of Vietnam to instruct commercial banks (especially state-owned commercial banks) to develop a preferential credit program with at least D100,000 billion for lending to high-tech and clean agriculture projects. The Ministry for Agriculture and Rural Development then provided criteria for identifying clean and high-tech agriculture projects. Subsequently, Agribank implemented a preferential credit program to support the production of clean and high-tech agriculture.[a] Target customers were businesses, cooperatives, cooperative unions, farm owners, etc., and interest rate reduction was about 0.5%–1.5% per year. Agribank also developed guidelines on environmental risk management in 2023 and incorporated environmental issues in its guidelines on credit extension and related procedures.

The prerequisites for effective implementation of this approach are (i) the capacity of state-owned banks to establish and execute the new lending processes (e.g., project identification, screening, risk assessment); and (ii) availability of technical assistance to support financial institution to adopt ESG into banking practice (e.g., incorporating ESG in business strategy, setting up a steering committee and a working group for ESG implementation, developing policies and procedures on environmental and social risk management in the credit activities).

Agribank enabled fast-growing green lending, yet there is a room for further growth. Green loan balances increased rapidly at 100%–380% in 2018–2020 with the total outstanding green loans amounting from D1.7 trillion ($68 million) to D12 trillion ($471 million), mainly for sustainable forestry, renewable and clean energy, and green agriculture.[b] As of March 2024, the total outstanding green loans amounted to $1.2 billion including $353 million for sustainable forestry, $612 million for renewable and clean energy, and $238 million for green agriculture.[c] Nevertheless, the bank's green credit accounted to about 1.8% of total loan book, and the green credit for the whole banking sector in Viet Nam is still at an early stage of development with a limited portfolio of around 4.5% of total loans. The Asian Development Bank (ADB) is providing technical assistance to selected banks to strengthen capacity for greener lending portfolio.

Good practice elements and discussion on applicability:

- **Set clear definition of "green" projects.**
- **Institute a robust risk management system.**

---

[a] Vietnam News Agency. 2023. *Agribank Invests in Hi-tech Agriculture*. 28 March.
[b] Nguyễn Thị Thu Hà. 2023. Green Banking in Agribank's Development Strategy. *Environment*. No. 4. pp. 41–45.
[c] Communist Party of Viet Nam. 2024. *Green Economy Investment Credit at AGRIBANK*. 7 August.
Source: ADB.

Box A23: Thailand's Government Savings Bank's Social Bonds

Thailand aims to achieve a just climate transition to address the challenges faced by vulnerable communities due to climate change. The Government Savings Bank (GSB) is a state-owned bank under the supervision of the Ministry of Finance. It aims to improve access to low-cost funding to support the grassroots population, small businesses, and local communities.[a] The climate transition will inevitably pose challenges to vulnerable communities, and investment in social infrastructure (e.g., retraining workers) is necessary. Although GSB does not explicitly target climate-related social investment, its approach could be leveraged to ensure a just transition.

The GSB established a social finance framework for its social bonds program, with technical assistance provided by the Asian Bond Markets Initiative and ASEAN Catalytic Green Finance Facility (ACGF) of the Asian Development Bank (ADB).[b] The framework included the project evaluation and selection process, management of proceeds, and periodic reporting and review. The proceeds were used for the following:

- Providing low-interest loans to the grassroots population to improve their living conditions.
- Supporting upskilling, especially among the unemployed and vulnerable groups impacted by the coronavirus disease (COVID-19) pandemic.
- Supporting small and medium-sized enterprises affected by the pandemic.

To ensure proper use of social bond proceeds, loan applications would be screened for eligibility in line with credit policies. A social bond working group was established to assess the existing lending portfolio and prospective projects against eligibility criteria. The shortlisted eligible projects would be submitted to the GSB president and chief executive officer for formal approval for proceeds allocation.

The prerequisites for the effective implementation of this approach are (i) the capacity of state-owned banks to establish and execute the new borrowing and lending processes, (ii) technical support from multilateral agencies for framework design and transaction execution, and (iii) investor appetite for these instruments.

The initiatives raised $295 million in 2022 issuance and demonstrated a case for other financial institutions and state-owned enterprises on how innovative debt instruments could be used to fund social investments.[c]

Good practice elements and discussion on applicability:

- **Test innovative debt instruments for raising public finance.**
- **Establish a robust governance and disclosure structure.**

---

[a] Government of Thailand, Government Savings Bank. 2022. *Social Finance Framework*.
[b] ADB. 2022. ADB Supports Thailand's First Social Bond Issued by State-Owned Government Savings Bank. News Release. 29 June.
[c] ADB. 2024. *Social Bonds for the Community Development in Thailand*. Video. 8 April.
Source: ADB.

Box A24: Singapore Sovereign Wealth Fund Advancing Climate Initiatives

GIC is a fund manager for the Government of Singapore. The Ministry of Finance gives GIC an investment mandate that stipulates the terms of appointment, investment objectives, risk parameters, investment horizon, and guidelines for managing the reserves. The finance minister sits on GIC's Investment Strategies Committee. As an investment vehicle with substantial resources, GIC aims to advance climate initiatives and achieve superior long-term returns by focusing on low-carbon and climate-resilient investment strategies.[a]

GIC is guided by its framework for sustainability:[b]

- **Capturing opportunities.** Allocate direct capital toward green solutions or enablers of the low-carbon transition via the Sustainable Investment Fund (SIF).

- **Protecting our portfolio.** Manage risks posed by assets that face high-stranding risk and support transition efforts by companies, with risk screening, robust due diligence, and portfolio stress testing.

- **Developing enterprise excellence and partnerships.** Manage GIC's carbon footprint and engage in research, content, and event partnerships to enhance the integration of climate science to address investment needs.

GIC has set up dedicated teams to capture sustainability-related investment opportunities:

- **Sustainability solutions group.** This is a dedicated team within private equity to deepen exposure to early-stage energy transition opportunities.

- **Transition and sustainable finance group.** This is a dedicated team within fixed income and multi-assets to invest in sustainability-related opportunities.

GIC also supports Task Force on Climate-related Financial Disclosures' recommendations and reports climate-related information on governance, strategy, risk management, and metrics and targets.

The prerequisites for the effective implementation of this approach are (i) buy-in from stakeholders (e.g., board of directors consisting of line ministers) and (ii) the capacity of sovereign wealth fund (SWF) management to execute the green mandate.

The initiative created significant impact through investment in climate tech enablers and sustainability data ecosystems. GIC has invested in Form Energy, which develops multiday energy storage solutions, and Climeworks, which aims to scale up direct air capture solutions. GIC has also invested in a supply chain sustainability rating company.

Good practice elements and discussion on applicability:

- **Establish clear investment guidelines.**
- **Dedicate resources in strengthening the local green investment ecosystem.**
- **Institute a robust climate disclosure system.**

---

[a] GIC. GIC: Our Governance.
[b] GIC. 2023. *FY2022/2023 Report on the Management of the Government's Portfolio.*
Source: ADB.

Mobilize and Deploy Private Capital and International Public Funds

Box A25: SDG Indonesia One Blended-Finance Platform

Indonesia needed to mobilize funding to bridge the funding gap to achieve the Sustainable Development Goals (SDGs). Given Indonesia's economic growth, the Government of Indonesia recognized that traditional donor development assistance would no longer be viable. Blended finance was identified as an effective tool to attract both domestic and international private capital to bridge the funding gap. In 2018, the Ministry of Finance and PT Sarana Multi Infrastruktur (PT SMI) launched SDG Indonesia One (SIO), a blended finance platform to channel private and public finance to achieve the SDGs.[a]

SIO partnered with 35 institutions for accessing capital and technical assistance to launch four different types of facilities to de-risk projects and increase bankability:

- development facilities, in partnership with the Asian Development Bank (ADB), KfW, the Asian Infrastructure Investment Bank, etc., to provide grants for project preparation and technical assistance to reduce the risk profile of projects;

- de-risking facilities, in partnership with AFD, Mentari, etc., to provide guarantees, investment premiums, etc. to reduce financial risk and enhance the credit profiles of projects;

- financing facilities in partnership with Standard Chartered, the European Investment Bank, etc. to provide senior loans and subordinated loans to mobilize commercial capital in the form of direct loans, co-financing, and loan syndication opportunities; and

- equity fund, in partnership with Engie, the China Communications Construction Company, etc., to provide equity and equity-linked investment to mobilize equity investment.

SIO consolidated different types of facilities in a central platform, which allows efficient matching between projects and targeted facilities to ensure the effective use of donor capital. SIO also follows the Organisation for Economic Co-operation and Development blended-finance principles as it (i) links to national priority sectors; (ii) integrates environmental, social, and governance factors in project selection; (iii) understands the mandates, objectives, and return profiles of partners; and (iv) makes sufficient funding available for early-stage project preparation.

The prerequisites for the effective implementation of this approach are (i) expertise to structure transactions on a deal-by-deal basis; (ii) the availability of different forms of risk-tolerant capital (e.g., direct grant, concessionary lending from development banks, equity investment from sovereign wealth fund); (iii) financiers' appetite to direct capital to the country; (iv) institutional capacity of domestic coordinating institution (PM SMI in this case) to engage different donors and project developers on different types of projects; and (v) applicability of international standards given the local context.

continued on next page

Box A25 continued

There have been some implementation challenges (e.g., identifying qualified subprojects that meet the necessary criteria), the initiative created a significant impact:[b]

- **Mobilized significant funding.** As of December 2022, the \$3.19 billion commitment was reached, of which around 64% was earmarked for climate change; 62 blended finance projects have been funded or developed under the SIO platform.

- **Enabling ecosystem activities.** In addition to financing, SIO conducted 46 capacity-building, knowledge-sharing, and business-matching activities.

- **Implementation models of blended finance structure.** Five different blended finance structures were implemented in projects (e.g., airport rooftop solar), building experience and capacity to scale up the approach.

Good practice elements and discussion on applicability:

- **Deploy risk-tolerant capital.**
- **Engage with different types of capital providers.**
- **Collaborate with IFIs, MDBs, and international partnerships.**
- **Harmonize the approach with regional and international standards**

---

[a] PT Sarana Multi Infrastruktur (Persero) - PT SMI. n.d.; and Government of Indonesia, Ministry of National Development Planning (BAPPENAS) and the United Nations Development Programme. 2022. Indonesia Integrated National Financing Framework 2022.

[b] F. Pranawa and PT SMI. 2022. SDG Indonesia One Blended Finance Platform. Presentation at Tri Hita Karana Blended Finance Deep Dive Workshop. 19 April.

Source: ADB.

Box A26: "FAST-P" Blended Finance Partnership

Asia makes up about half of global greenhouse gas (GHG) emissions, however many transition projects remain unviable to commercial capital due to political, currency, and credit risk. Bridging the financing gap requires innovative financial instruments and new partnerships. Blended finance can unlock capital needed to achieve Asia's energy transition by reducing the cost of capital for energy transition projects.

A blended finance initiative, Financing Asia's Transition Partnership (FAST-P), established by Monetary Authority of Singapore (MAS) aims to mobilize capital for climate projects in Asia. FAST-P is set up in Singapore through collaboration between key public, private, and philanthropic sector partners. It aims to mobilize up to $5 billion to de-risk and finance transition and marginally bankable green projects in Asia. It brings together diverse stakeholders such as multilateral development banks, sovereign partners, philanthropic organizations, and the financial sector to support Asia's decarbonization and narrow the financing gap. Singapore is committed to contributing concessional capital with other partners crowding in commercial capital.[a]

FAST-P will address three critical areas of climate action-related investment:

- **Energy transition acceleration.** To support projects like the managed phase-out of coal-fired power plants and their replacement with renewable energy sources.

- **Green investments.** To focus on mature technologies such as renewable energy scaling, grid modernization, and electric vehicle infrastructure.

- **Clean technologies.** to invest in emerging green technologies like hydrogen and carbon capture, utilization, and storage.

The prerequisites for the effective implementation of this approach are (i) availability of concessional finance and grants from multilateral development banks and international financial institutions; (ii) stable and supportive policy and regulatory environment; (iii) financiers' appetite to participate in the innovative financing structure; and (iv) institutional capacity of a coordinating institution (e.g., MAS) to engage different donors and stakeholders on different types of projects.

Good practice elements and discussion on applicability:

- **Deploy risk-tolerant capital.**
- **Engage with different types of capital providers.**
- **Collaborate with IFIs, MDBs, and international partnerships.**

---

[a]   Asian Development Bank (ADB). 2023. *ADB, GEAPP, and MAS to Establish Energy Transition Acceleration Finance Partnership in Asia*. News Release. 5 December; and Monetary Authority of Singapore (MAS). 2024. *Financing Asia's Transition – Briding Global Ambition to Regional Action and Impact*. Presentation at the Financing Asia's Transition (FAST) Conference. 17 April.
Source: ADB.

Box A27: Blended Finance for Cambodia's National Solar Park

Cambodia faces the challenge of increasing renewables in its energy mix while ensuring the low cost of power generation. In 2018, the Asian Development Bank (ADB) partnered with Electricite du Cambodge (EDC), Cambodia's national power utility, to develop a 100-megawatt (MW) National Solar Park including a transmission interconnection system. The project also strengthened capacity of EDC and Electricity Authority of Cambodia (national electricity regulator) for integrating renewable energy into the national grid.

This project demonstrated the potential to develop large solar photovoltaic (PV) in a cost-effective manner in Cambodia by utilizing a combination of concessional loans and grant resources. The project combined capital from various sources:

- $11 million loan and $3 million grant from the Strategic Climate Fund (under the World Bank);
- $7.6 million concessional loan from ADB;
- $0.5 million from the Republic of Korea e-Asia and Knowledge Partnership Fund to support the EDC's capacity development.[a]

ADB provided end-to-end technical assistance for the transaction:[b]

- The project was prepared with grants from the governments of Canada and Singapore.
- ADB's Office of Public–Private Partnership assisted with the design and execution of the competitive tender. It also advised on the transaction, covering project due diligence, market sounding, feasibility study preparation, etc.
- ADB helped access cofinancing for the project through a combination of concessional loan and grant resources from the Strategic Climate Fund.

The prerequisites for the effective implementation of this approach are (i) the availability of concessional finance and grants from multilateral development banks and international financial institutions, (ii) a stable and supportive policy environment, and (iii) the private sector's capacity to participate in tenders and offer services.

The project delivered a record-low procurement price. In 2022, the project reached a milestone with the park's first 60MW solar PV power generation plant connecting to the national grid. Phase I achieved a $0.039 per kilowatt-hour (kWh) cost while Phase II achieved $0.026 per kWh; both were record-low prices for utility-scale, grid-connected solar PV in Southeast Asia. ADB also has transaction advisory services with the EDC to help develop 2 gigawatts of solar power (footnote b).

Good practice elements and discussion on applicability:

- **Engage with different types of capital providers.**
- **Collaborate with IFIs, MDBs, and international partnerships.**

---

[a] ASEAN Energy Database System. 2018. ADB Boosts Cambodia's Energy with $7.6mln for Solar Park. News release.
[b] ADB. 2022. ADB-supported National Solar Park in Cambodia Connects to Grid. 15 November.
Source: ADB.

Box A28: ASEAN Catalytic Green Finance Facility

The Association of Southeast Asian Nations (ASEAN) member states need significant investment, around $210 billion per year, for infrastructure development, especially green infrastructure projects. Creating bankable green investments requires private capital to play a much larger role and public finance to be used strategically to crowd in private sources of green finance and investment. For green private capital flows to scale up rapidly, there is a need to support commercially feasible projects.

In 2019, the ASEAN Catalytic Green Finance Facility (ACGF) was launched to accelerate green infrastructure investments in Southeast Asia. The ACGF is owned by ASEAN governments and the Asian Development Bank (ADB), managed by ADB's Southeast Asia Green Finance Hub, and supported by nine co-financing partners: ADB, Agence Française de Développement, Cassa Depositi e Prestiti, European Investment Bank, the European Union, Green Climate Fund, Government of the Republic of Korea, KfW, and the Government of the United Kingdom.

It helps bridge the infrastructure investment gap by supporting the identification and preparation of commercially viable green infrastructure projects:[a]

- **Origination.** Potential projects for early stage support identified from ACGF country roundtable events and consultations with ADB operations teams and ACGF partners.

- **Rapid assessment.** Technical support provided for initial project preparation, including financial structuring and an assessment of potential green impacts.

- **Screening and funding interest.** Projects were screened against ACGF eligibility criteria, and a draft project information memorandum was produced. Financing interest was sought from ACGF cofinancing partners.

- **ASEAN Infrastructure Fund Board review.** A full project information memorandum (including a financing plan) was finalized and submitted to the ASEAN Infrastructure Fund Board for review and approval.

- **Project preparation and loan approval.** Due diligence was conducted, and the project was prepared for ADB and ACGF co-financing partner loan approvals.

The ACGF also provides knowledge and training programs to strengthen the regulatory environment and build the institutional capacity of ASEAN governments to scale up green infrastructure investments.

The prerequisites for the effective implementation of this approach are (i) a pipeline of high-quality green investment projects that meet the eligibility criteria to ensure there are sufficient eligible projects to be funded; (ii) financiers' appetite to participate in the innovative co-financing structure; and (iii) the availability of concessional finance and grants from multilateral development banks and international financial institutions.

The ACGF mobilized significant funding for green infrastructure projects in ASEAN member states. Nine partners committed $1.8 billion in cofinancing and funds for technical assistance to the ACGF. By the end of 2022, $504 million of the partner funds were committed to ACGF-eligible projects in the ADB pipeline.

Good practice elements and discussion on applicability:

- **Deploy risk-tolerant capital.**
- **Engage with different types of capital providers.**
- **Collaborate with IFIs, MDBs, and international partnerships.**

---

[a] ADB. ASEAN Catalytic Green Finance Facility.
Source: ADB.

Box A29: Indonesia's Green Taxonomy

To facilitate green finance, a harmonized understanding of terminologies and criteria across companies, financial institutions, investors, policymakers, and other relevant stakeholders is required. Many green taxonomies are being developed across the world (e.g., Association of Southeast Asian Nations [ASEAN] Taxonomy, European Union [EU] Taxonomy), and it is challenging to ensure consistency with and interoperability across these. Moreover, it proves challenging to balance stringent standards with practical implementation. While a basic traffic light system may lack credibility, a taxonomy with detailed screening criteria and thresholds for each specific activity may pose difficulties in adoption.[a]

Indonesia's Taxonomy 1.0 adopts the Foundation Framework from the ASEAN Taxonomy.[b] It classifies and color-codes activities based on their impact on the environment and on climate mitigation and adaptation.

- green is for activities with no significant harm, a positive environmental impact, and that align with national environmental objectives;

- yellow is for activities with no significant harm, but that are still transitioning to align with national environmental objectives; and

- red is for environmentally harmful activities.

In 2024, the Financial Services Authority of Indonesia published Taxonomy 2.0, which had been updated to be interoperable with other taxonomies. A Sustainable Finance Task Force was established, which brought together over 47 financial market players including commercial and Islamic banks, issuers, securities companies, and investment managers. The purpose of this task force is to collaboratively develop policies and support capacity-building initiatives for private financial institutions. In addition, feedback was gathered from academics, nongovernment organizations, international organizations, etc.[c]

The prerequisites for the effective implementation of this approach are (i) close collaboration across line ministries to define activities and align on standards and requirements, (ii) maturity of the financial market and market players to support the development of a national green taxonomy, and (iii) the applicability of international standards given local context.

The inclusion of coal sparked discussions on Green Taxonomy 2.0. Investments in captive coal plants can be considered as "transition" activities if the plants commit to cut greenhouse gas emissions by 35% within 10 years and by 100% by 2050. Financial Services Authority chief Mahendra Sirega responded that the new taxonomy incorporated a broader set of considerations, "not just the environmental aspect, but also balancing that with social progress and economic development aspects" (footnote c).

Good practice elements and discussion on applicability:

- **Harmonize approach with regional and international standards.**
- **Conduct stakeholder engagement and foster collaboration.**

---

[a] Coalition of Finance Ministers for Climate Action (CFMCA). 2023. *Strengthening the Role of Ministries of Finance in Driving Climate Action. A Framework and Guide for Ministers and Ministries of Finance.*
[b] L. K. Larasati and T. Mafira. 2022. Indonesia Green Taxonomy 1.0: Yellow Does Not Mean Go. *Climate Policy Initiative Blog.* 22 July.
[c] K. Lee and O. J. Keuangan. 2024. Indonesia Sparks Criticism with Role for Coal in Green Taxonomy. *Green Central Banking.* 26 February.
Source: ADB.

Box A30: Singapore's Grant Scheme to Support Green Loan Verification and Lending Framework Development

Climate finance instruments by banks, such as sustainability-linked loans, are a key source of financing for climate action of corporates, small and medium-sized enterprise (SMEs), and households. However, there are additional costs associated with these instruments, both for the lender and borrower. Banks will need to develop frameworks that delineate clear criteria for evaluating prospective green and sustainable finance transactions. Borrowers will incur costs for verification and reporting to satisfy the lending criteria. Government grants that cover these additional costs could accelerate the offering and adoption of green loans.

The Monetary Authority of Singapore (MAS) launched the Green and Sustainability-Linked Loan Grant Scheme (GSLS) starting in 2021.[a] The scheme incentivizes the provision and application of green and sustainability-linked loans:

- For borrowers, the GSLS offers up to S$ 100,000 per loan over 3 years to cover expenses associated with verification pre-borrowing and reporting post-borrowing.

- For banks, GSLS covers up to S$ 180,000 on expenses incurred to develop GSLS frameworks for SMEs and individuals; and up to S$ 120,000 for other lending.

The GSLS is designed to build up the ecosystem for scaling the adoption of climate finance. Specifically, by creating demand for assessment, verification, and reporting services, it supports the development of the domestic enabling market infrastructure:

- **Service provider requirement.** Expenses are covered by GSLS only if the verification or framework advisory service provider has more than 50% of gross revenue attributable to Singapore.

- **Bank requirement.** Expenses are covered by GSLS only if the design and conceptualization of the framework are performed in Singapore.

The prerequisites for the effective implementation of this approach are (i) the availability of high-quality local verification and reporting services; (ii) the institutional capacity of domestic banks to implement newly developed loan frameworks; (iii) domestic talent pool and business ecosystem for provision of ancillary services (e.g., consulting, climate data) related to climate finance; and (iv) high and sustained future demand for climate finance and ancillary services.

The volume of GSLS grew to S$17 billion in 2021 (+60% from 2020). Banks were incentivized to develop green loan frameworks. Accompanying the launch of the GSLS, BNP Paribas, OCBC Bank, and UOB have introduced innovative GSLS frameworks that will qualify for the scheme.[b]

Good practice element and discussion on applicability:

- **Support capacity-building for lenders and borrowers of climate investment.**

---

[a] MAS. 2020. *MAS Launches World's First Grant Scheme to Support Green and Sustainability-Linked Loans.* Media Release. 24 November; and MAS. 2020. *Green & Sustainability-Linked Loan Grant Scheme (GSLS) Brochure.*
[b] MAS. Schemes and Initiatives. Sustainable Loan Grant Scheme.
Source: ADB.